STP 1493

Repair, Retrofit and Inspection of Building Exterior Wall Systems

Paul G. Johnson and Jon M. Boyd, editors

ASTM Stock Number: STP1493

ASTM International
100 Barr Harbor Drive
PO Box C700
West Conshohocken, PA 19428-2959

INTERNATIONAL
Standards Worldwide

Printed in the U.S.A.

Library of Congress Cataloging-in-Publication Data

Repair, retrofit, and inspection of building exterior wall systems / Paul G. Johnson and Jon M. Boyd, editors.
p. cm.
Includes bibliographical references and indexes.
"ASTM Stock Number: STP1493."
ISBN 978-0-8031-3418-8
1. Exterior walls--Maintenance and repair. I. Johnson, Paul G., 1949- II. Boyd, Jon M., 1952-

TH2235.R46 2009
690'.120288--dc22

2009003037

ISBN: 978-0-8031-3418-8

Peer Review Policy

Each paper published in this volume was evaluated by two peer reviewers and at least one editor. The authors addressed all of the reviewers' comments to the satisfaction of both the technical editor(s) and the ASTM International Committee on Publications.

The quality of the papers in this publication reflects not only the obvious efforts of the authors and the technical editor(s), but also the work of the peer reviewers. In keeping with long-standing publication practices, ASTM International maintains the anonymity of the peer reviewers. The ASTM International Committee on Publications acknowledges with appreciation their dedication and contribution of time and effort on behalf of ASTM International.

Citation of Papers

When citing papers from this publication, the appropriate citation includes the paper author(s), "paper title", J. ASTM Intl., volume and number, ASTM International, West Conshohocken, PA, Paper, year (listed in the footnote to the paper), Paper ID JAI100887 (located at the top of each first page). This book is a repackaging of those papers.

Printed in Dayton, OH
March 2009

Foreword

This publication, *Repair, Retrofit and Inspection of Building Exterior Wall Systems*, contains papers presented at the ASTM symposium of the same name held on 21, 22 October, 2006 in Atlanta, GA. The symposium was presented by ASTM Committee E-6 on Performance of Buildings. The symposium co-chairmen were: Jon Boyd, Klein and Hoffman, Inc., Chicago, Illinois and Paul Johnson, Smith Group, Detroit, Michigan.

Contents

Overview

The papers published in this special technical publication were presented during the ASTM symposium entitled *Repair, Retrofit and Inspection of Building Exterior Wall Systems*, held in October 2006 in Atlanta, GA, as part of a continuing series of symposia on building exterior walls system presented by the sponsoring subcommittee, ASTM E06.55 on Building Exterior Wall System.

The 2006 symposium continues the work began by E06.55 back in 1990 to bring the talents and diverse interests of the committee and the building industry together to exchange information and experience regarding the building envelope. It continues to be the goal of the committee to address the complex issues of design, construction, maintenance, evaluation and repair of these important systems of our buildings. With a very large stock of existing buildings we will be faced with the special problems and needs of these structures as they age, and as we ask their exterior wall systems to perform in new ways for us. This was the driving factor behind this symposium and many of the resulting papers.

The papers presented in this symposium addressed a diverse range of topics including survey and assessment techniques and sampling, establishing the integrity of cladding systems and components, case studies of system deterioration, and remediation, serviceability issues, seismic retrofit, energy savings attributable to façade upgrades, and re-cladding of existing buildings.

System types addressed include façades comprised of vintage masonry, limestone, precast concrete panels, thin stone, exposed cast-in-place concrete, aluminum framing, EIFS, and glass and metal walls.

These papers represent a broad range of experience and perspectives of the authors arising from varying backgrounds and experience, professions, and geographic locations.

It is our hope that this publication and others from this subcommittee will offer meaningful real-world insights into the complex and challenging problems associated with the design, construction and maintenance of building exterior wall systems, both old and new.

Jon M. Boyd, SE, AIA
Klein and Hoffman, Inc.
Chicago, Illinois

Paul G. Johnson, FAIA
The Smith Group
Detroit, Michigan

TESTING AND EVALUATION TECHNIQUES

Journal of ASTM International, Vol. 4, No. 8
Paper ID JAI100809
Available online at www.astm.org

Kevin S. Coll[1] *and David A. VanOcker*[1]

Case History of the Use of Electronic Survey Techniques to Assess Planar Distortions in Building Facades

ABSTRACT: The occurrence of distortions within facade elements in the form of leaning, bowed and vertically displaced wall construction is common in older structures. Quite often these conditions are identified by building owners and serve as the primary focus of investigations to assess their significance. Quantifying the severity and extent of these displacements is integral to these evaluations. Various surveying methods are available to assist the investigator in determining planar distortions, ranging from simple plumb bob and level surveys to digital electronic surveying techniques. This paper focuses on the use of electronic surveying techniques as a diagnostic tool in assessing planar distortions and displacement in three dimensions on building facades. A case study is presented as a typical application where planar displacements within a building facade needed to be assessed. To provide context for selection, use, and interpretation of various survey techniques, a project is reviewed in detail from initial identification of displacements, review/selection of an appropriate survey technique, manipulation, and interpretation of survey data, assessing the significance of the data relative to the case study building, and review of repair program parameters.

KEYWORDS: facade, survey, assessment, investigation, masonry, planar, distortion, displacement, building envelope, electronic survey techniques

Introduction

Electronic surveying techniques have proven to be useful means of gathering large quantities of displacement data in a timely and cost-effective manner when access restrictions and scheduling pressures make more conventional surveying techniques unfeasible. Situations in which the quantification of planar distortions and displacements have proven to be necessary have consisted of the evaluation of aging stone masonry structures for bowing/leaning associated with a combination of deterioration related to (1) long-term water penetration into the wall system, (2) damage related to a specific incident (i.e., fire, snow storm), (3) displacements related to oxide corrosion of embedded steel components, (4) corrosion of wall ties/stone anchors resulting in loss of veneer anchorage, and (5) differential settlement, or any combination of the above. The following case study is presented as a representative project in which this survey technique was successfully employed.

The selected subject building is a collegiate dormitory constructed circa 1890, located in the northeastern United States (see Fig. 1). The predominant building construction type is multi-wythe unreinforced masonry bearing walls, comprised of a schist stone veneer built compositely with rubble stone backup, supporting wood floor and roof framing. Interestingly, the original basis of our building investigation was to confirm the scope of what was expected to be a minor facade masonry repair program, ancillary to a larger roofing project to address significant roof leaks. Unfortunately, as is so often the case with historic structures, our investigation expanded significantly to assess apparent out-of-plane distortions in the walls that were found to correspond to considerable hidden structural distress conditions.

Preliminary Investigation and Analysis

The first component of our assessment consisted of a document review of the institution's archives to research available records for details on the building construction. A partial drawing set dating to the original construction was obtained along with project documents from various renovation programs. A

Manuscript received September 22, 2006; accepted for publication July 18, 2007; published online August 2007.
[1] Project Manager and Principal, respectively, CVM Facilities Renewal, 85 Old Eagle School Road, Wayne, PA 19087.

FIG. 1—*Photo of building shortly after original construction (circa 1890)*

review of the College's photographic archive revealed that the subject building experienced a major fire that affected approximately 80 % of its footprint, after which the building was reconstructed using new wooden framing supported on the original masonry bearing walls.

Our preliminary visual survey of the building exterior identified significantly bulged and/or leaning portions of the exterior stone masonry walls. These distortions appeared to be on the order of several inches over the height of one story.

We initially concluded that the conditions appeared to be of sufficient magnitude to present a potential life-safety hazard to the building occupants and pedestrians. A detailed survey was necessary to quantify the extent and severity of the distortions in order to enable an informed discussion of the conditions with the owner. The severity of the conditions raised the issue of limiting use of the building and/or implementing temporary stabilization measures. Given the building occupancy and scheduling constraints of installing any immediate stabilization measures, the detailed assessment of the degree and nature of the planar displacement had to be performed as quickly as possible.

Detailed Surveying Techniques Considered

Several surveying methods were considered as potential options to establish the relative displacements within the building envelope, including plumb bob/line level surveys, vertically oriented laser level surveys, hand-held electronic distance measurement (EDM) device surveys, and various point and surface electronic surveying methods. Each method has relative benefits and restrictions, as tabulated in Table 1. The parameters of this project required that the entire building be surveyed within a one-week timeframe. Direct access to the building exterior was limited by its proximity to public streets, dense vegetation around areas of the perimeter, and an adjacent underground structure. With these parameters, the most cost-effective technique was determined to be the pulse-laser surveying method. This method can obtain large amounts of data quickly, does not require direct access to the building element being measured, provides measurements in three dimensions to a high degree of precision (5 mm), and provides data in standard electronic formats for interpretation/record.

TABLE 1—*Summary of survey method options considered.*

Method	Access Needed	Measures Displacement: Horizontal	Measures Displacement: Vertical	Labor	Comments
Visual	No	No	No	Minimal	Does not quantify degree of displacement
Straight Level	Yes	Yes	Yes	High	Access needed; time consuming; difficult to extrapolate data over large areas; minimal precision
Line Level/Plumb Bob	Yes	Yes	Yes	High	Access needed; time consuming; difficult to extrapolate data over large areas.
Laser Level (Vertical)	Yes	Yes	No	Moderate	Access needed; measures only horizontal displacements
Hand Held EDM (Range-Finder)	No	Yes	Yes	High	Less precise than than other EDM methods; extensive manual calculation needed to interpret data
Pulse-Laser Surveying	No	Yes	Yes	Minimal	Most precise of EDM methods considered; data is provided in electronic format that eases interpretation
Infared Laser Surface Scan	No	Yes	Yes	Minimal	Not as precise as pulse laser method; equipment not readily available in some areas

Pulse-Laser Surveying Methodology

An electronic survey was conducted on all elevations of the building using NIKON NPL Series pulse-laser total station survey equipment. The surveying methodology utilized in acquiring the data followed standard surveying practices (i.e., establishing survey stations, back-shooting reference points and benchmarks to verify precise location in three dimensions, etc.). To facilitate the interpretation of the data, survey points were referenced along local North-South and East-West axes established to correlate with the orientation of the primary building walls. The parameters of the surveying program were established at a presurvey conference held in the field with the surveyor, identifying specific building components to shoot in a systematic and repeatable fashion. The goal was to obtain sufficiently arrayed and detailed information on the building facade to enable our assessment. At this conference, the precise scope of the survey was established on a point by point basis for the entire facade.

For the case study project, the sandstone window surrounds were generally utilized as the survey points, as a majority of the displacements were observed at these elements and further, they had a consistent profile relative to the vertical and horizontal plane of the wall. The variable profile of the ashlar stone comprising the main fields of the walls could not serve as a meaningful reference ($\pm$2-in. amplitude). In total, over 1000 data points were obtained over a two-day period.

Analysis of Electronic Survey Data

Raw survey data were provided in electronic spreadsheet format along with keyed elevations locating the individual survey points on the building. Our preliminary analysis of the data involved hand-calculating local displacements on isolated elements of the building. It was during this analysis that we realized the importance of selecting key data points to utilize as local reference points for a given region. Ideally, reference points would represent the original position of the wall and would serve as a baseline for comparison for other readings. In the subject building, this process was complicated as the walls were already significantly displaced and the question arose as to what represented the original plane of each wall. In the end, several reference points were selected and back-checked against each other to establish what resulted in the most meaningful representation of the data. Typically, these reference points were selected in areas where the walls had additional bracing support (i.e., corners and intersections, shear wall interfaces, at floor diaphragms, etc.), which limited or prevented distortion in these regions.

Following this initial analysis of the survey results, the data was integrated into a three-dimensional computer-aided design (3D CAD) model to assess the extent of the displacement over the entire building facade. The full set of data points as incorporated into this model are denoted in Fig. 2 below, projected above the roof plan to help orient the user.

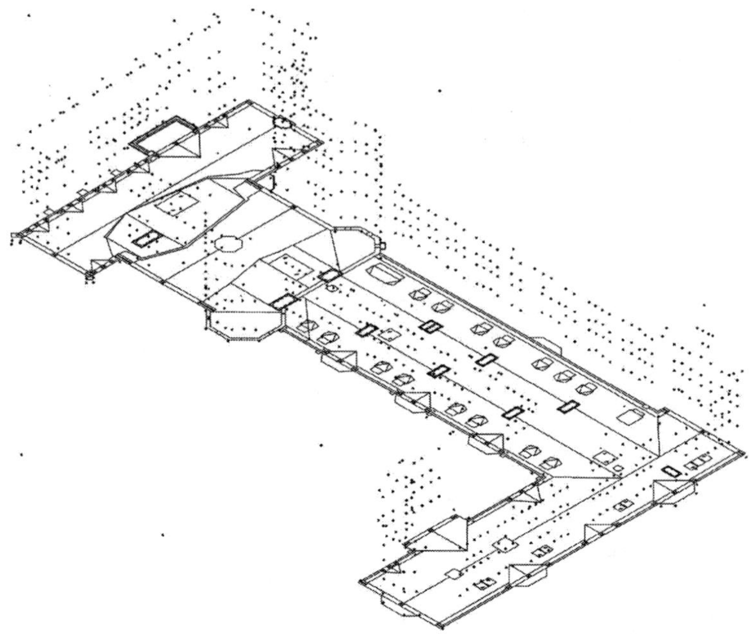

FIG. 2—*3D CAD model of survey data.*

To further facilitate the interpretation of the large amounts of data, a simple program was written to automate comparative calculations between individual and sets of points and a reference point selected at the user's discretion. Figure 3 portrays a portion of the building facade for which this technique was utilized. In this instance, the data points are layered over the building elevation for reference. The set of data points (portrayed as numbered circles) were selected for comparison to a reference point on the building (in this case a survey point located at the intersection of two walls). The CAD program then calculated the offsets in and out of the plane of the wall of the selected points relative to this reference point. Review of the information provided in Fig. 3 revealed that the top of the gable wall was leaning into the plane of the wall, while the head of the large window below the gable was bowed out of the plane of the wall. This process was repeated on individual wall segments for the remainder of the building.

Data analysis established that gable end walls and parapets were the predominant building elements exhibiting significant planar displacements. Figure 4 illustrates the typical extents of the localized displacements found on the building. The magnitude of displacement ranged from a maximum of 4 in. "out of" and 4 in. "in to" the original plane of the wall. A diagrammatic representation of the typical displacements observed at gable end walls is depicted in Fig. 5.

Building Probes of Existing Construction

Concurrently with the implementation of the electronic survey of the building facade, localized probes were conducted to establish the underlying cause(s) of the observed displacements. Surveys within unfinished attic spaces were supplemented with probes through interior finishes at other levels to verify the type and configuration of the wood framing. Additionally, isolated probes in the exterior masonry were conducted to confirm the masonry wall construction and specifically the type and condition of lintels supporting masonry above window openings (refer to Figs. 6 and 7 for typical lintel configuration).

The survey and probe investigation revealed that the wood floor and roof framing was typically pocketed into the exterior masonry wall construction without mechanical connections (e.g., metal strap, tie, or similar). This was a typical practice at the time of original construction; however, no positive means

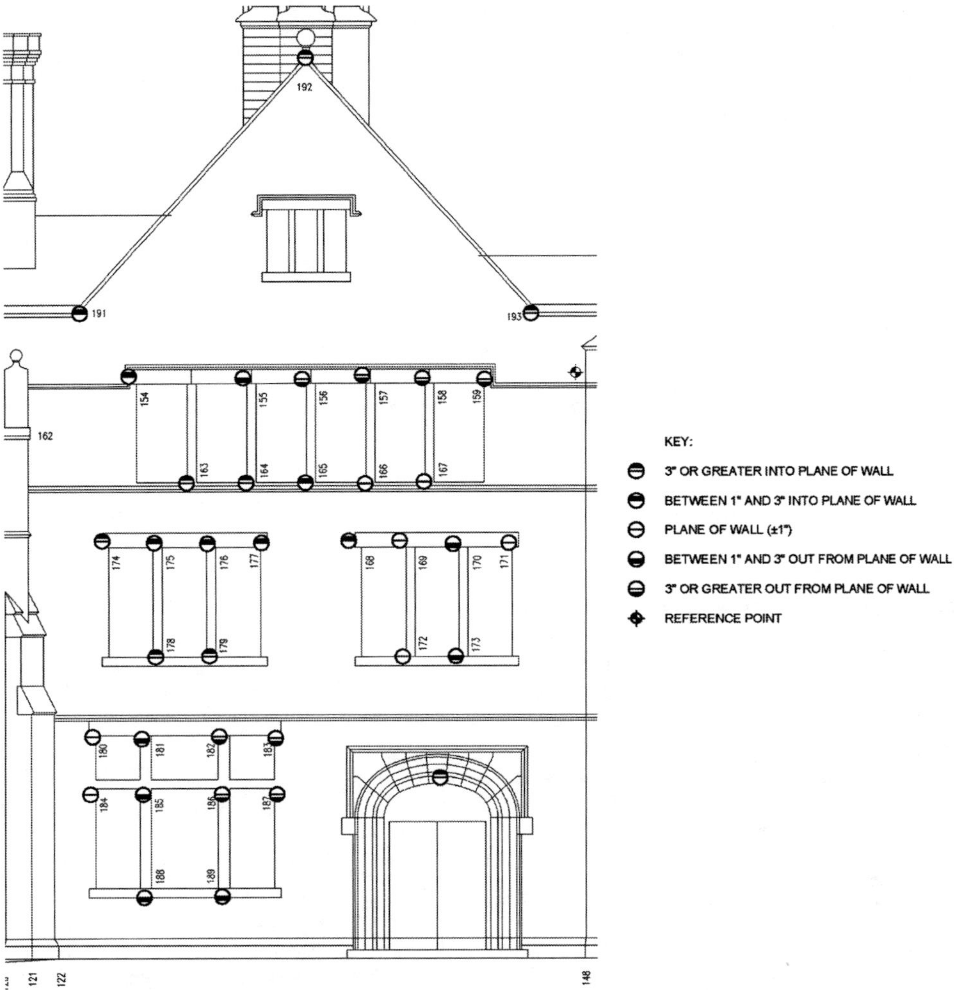

FIG. 3—*CAD Comparative analysis of typical wall area.*

FIG. 4—*Shaded regions of building elevation depicting areas of displaced masonry.*

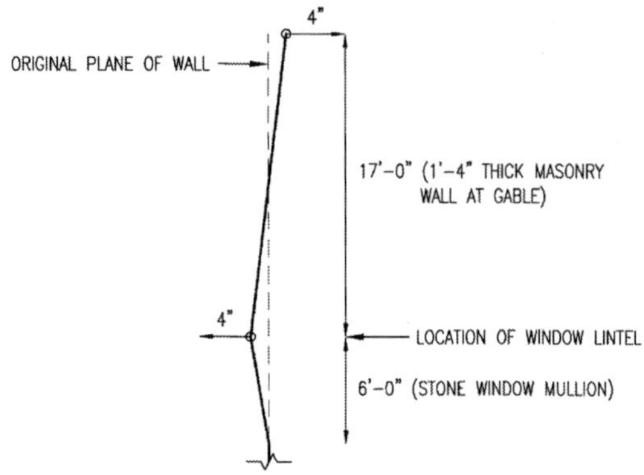

FIG. 5—*Diagrammatic representation of horizontal displacement at gable end wall.*

existed for the floors or roofs to provide any significant lateral bracing to the displaced exterior masonry walls.

The veneer masonry at window openings was supported by masonry elements (i.e., sandstone lintels and mullions), whereas the backup masonry was typically supported on a series of wood lintels. The only exceptions to this arrangement were two locations at which steel lintels were installed in lieu of wood above particularly large windows openings.

The backup rubble masonry was generally observed to be in fair condition considering the age of the building. Isolated areas in poor condition were encountered; these areas typically occurred at higher elevations on the building, under and adjacent to the built-in roof gutter system, and under parapet/gable coping stones. The generally poor state of the backup masonry in these areas was attributable to a history of prolonged exposure to water through defects in the water management system (i.e., failed sealant in coping joints, gutter/internal rain water conductor failures, lack of wall flashings, etc.). Interestingly, this correlated closely with the areas exhibiting the most significant displacement.

Finally, during the course of our survey and limited probing, deterioration of wood framing was observed in wood lintels, rafter and joist ends at masonry pockets, roof rafter ends, and rafter plates. Similar to the areas with deteriorated masonry backup, the locations where significant deterioration in the wood framing was observed correlated with areas where prolonged exposure to excessive water was believed to have occurred. The source of this water was generally found to relate to defects in the roofing system, in particular the built-in copper gutter system at through wall scupper and leader penetrations in the masonry parapet. Clearly defining the extent and severity of the deterioration in the wood framing proved to be difficult, as the amount of deterioration varied significantly within similar components in close proximity to each other (i.e., single joists containing significant deterioration were discovered within a series of sound joists). Visual inspection was also of limited use without systematic probing, as floor joist framing often appeared to be undamaged when viewed from the interior, but when probed to expose the full depth of the pocket, significant deterioration was discovered hidden within the plane of the wall.

FIG. 6—*View of exterior wall after partial demolition during repairs. Note the set of three wood lintels above window opening and deterioration wood floor and roof framing. The wood framing was temporarily supported by shoring inside the building (out of view of camera).*

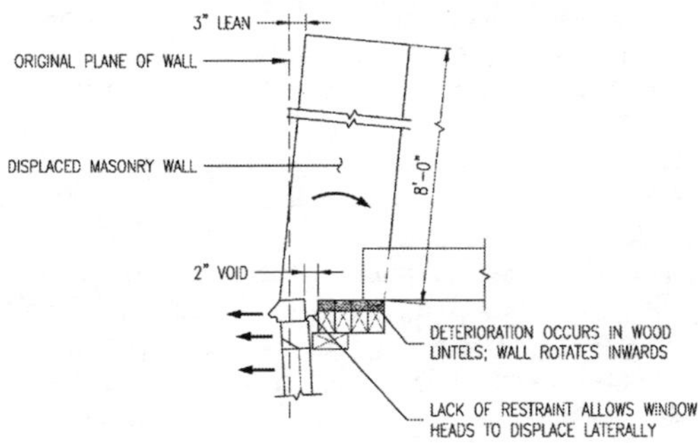

FIG. 7—*Wall section at typical existing window penetration.*

FIG. 8—*Photo of building after 1906 fire.*

Archival photographs indicate that the wood framing in the majority of the southern end of the building collapsed during the fire, along with localized portions of the exterior masonry walls, as shown in Fig. 8. The photographs reveal that the exterior walls and chimneys were left unbraced by the interior floor and roof framing for a period of time while the building was cleared of debris and reconstructed (note the corbelled chimneys on the right side of Fig. 8 and gable wall on the left, also portrayed in Figs. 3 and 4). Visual survey of the roof framing indicated that at least a portion of the displacements were already expressed within the structure at the time of the post-fire reconstruction, as evidenced by the roof framing being rebuilt to accommodate gable wall regions that were already leaning inward. While it is speculative to assess what contribution the fire may have had on the planar displacements of the walls, based on these observations we considered it likely that the fire played a role.

Structural Significance of Survey Data

Based on the detailed planar displacement survey and the information gathered on the structure utilizing more traditional survey and probing techniques, it was determined that the planar displacements were primarily related to the deterioration of and deficiencies in the lintel system utilized at wall penetrations. In all instances, the outer wythe of veneer masonry was supported on traditional stone masonry elements (i.e., lintels, mullions, etc.), while the backup masonry that comprised the majority of the wall thickness was supported on wood lintels and at isolated locations steel lintels. The outer and inner portions of the wall system were thus vulnerable to differential movement due to deterioration of the interior lintels. The lack of mechanical ties between the wood floor/roof framing, lintel system, and the exterior masonry, coupled with the generally poor condition of the masonry backup in the areas of most significant displacement, were also found to contribute to the displacements. The typical mechanism that led to the observed displacements is depicted in Fig. 7.

The development of the displacements appears to have occurred slowly over a period time, likely initiated as result of the 1906 fire, but progressing as deterioration of the wall components (i.e., lintels, backup masonry, etc.) occurred. Substantial evidence was available indicating that a significant portion of the displacement had occurred in the more recent past (as indicated by failed sealant joints installed in the 1980s) and that this movement was likely continuing. Consideration was given to quantifying the rate of movement within the building facade. However, given the nature and magnitude of the displacements, schedule pressures and the relative certainty that significant repairs would need to be implemented in a matter of months, it was decided that this additional information was unimportant to the overall project goal; the general scope of the repairs had been sufficiently established based on the data acquired.

Repair Program Implementation Strategy and Timeline

Within two weeks of commencing our preliminary visual survey, our assessment was sufficiently refined to enable a comprehensive discussion with the owner regarding the condition of the masonry walls and the parameters of necessary repairs. Based on this assessment and the period of time remaining until the students vacated the premise for the summer, it was determined that immediate stabilization measures and/or restrictions regarding the use of the building were not warranted. This was based on the understanding that a substantial repair effort would be implemented prior to the start of the fall semester.

Through continued discussions with the owner to refine the implementation of necessary repairs from a technical/logistic standpoint, concurrently with the owner's budget and scheduling constraints (i.e., the building needed to be fully occupied during the school year), it was decided to conduct the repairs in a phased manner during summer months over a three year period. Priority was given to address conditions that represented the more significant structural or life-safety hazards or both, during the first phase.

Selective demolition and complete reconstruction of the affected areas was judged to be the only feasible repair approach, considering the type of wall construction and the magnitude of the wall displacements, coupled with the poor condition of the backup masonry and wood lintels. All wood lintels in rebuild areas were replaced with precast concrete/masonry bond beam lintels. A redesign of the wood framing support system in all rebuild areas was considered to be another integral component of the repair program, eliminating the use of masonry pocket bearings. To accomplish this, new stainless steel joist hanger hardware was integrated into the masonry backup wall during reconstruction. This connection served the dual purpose of isolating the wood framing from the exterior masonry wall construction, minimizing moisture exposure, while enhancing the lateral tie connection between the wall system and interior framing.

The regions requiring repairs to the wood framing were known only broadly at the completion of the survey period. Fully documenting this component of the repair program prior to the start of construction was not feasible, considering as-built variations in framing and the difficulty in quantifying the repair locations without a lengthy assessment and numerous probes. Representative wood framing repairs were detailed and repair quantities estimated, with the final scope of repairs confirmed during construction as areas were fully exposed. A contingency accommodated additional repairs necessitated by unforeseen variations in the type and condition of the framing.

Concluding Remarks

The use of electronic surveying techniques to determine the presence and severity of planar displacements proved to be a cost-effective and essential diagnostic tool. In this case study, the use of the pulse-laser total station survey method enabled our assessment to reach conclusions as to the nature and extent of the distress in the wall system in a very short timeframe. This facilitated an informed discussion with the owner in which the safety of the structure in its current condition and parameters of necessary repairs were reviewed, and the owner's desire to expedite long-term budgetary and occupancy planning relative to the implementation of needed repairs was met.

Although determined not to be necessary for the case study project, the pulse-laser survey system is well suited to the repeated surveying of the same points over time as part of a displacement monitoring program. We have found this survey technology to be useful for recording active differential movement within a building system, such as that resulting from sinkhole activity or improperly executed underpinning. Such applications require that permanent monitoring points be established on the building to create reproducible data sets for comparison over time. Similarly, we have utilized this technique to assess the efficacy of subsurface remediation measures, and as a control measure to record and assess movement related to the implementation of potentially dynamic operations such as compaction grouting (which has the potential and is at times utilized to jack a displaced structure back into position.)

Although electronic survey techniques are considered initially more expensive, their speed of implementation and reproducibility can provide the most cost-effective solution for these more structurally intensive investigations.

Journal of ASTM International, Vol. 4, No. 9
Paper ID JAI100815
Available online at www.astm.org

Lonnie L. Haughton[1] *and Colin R. Murphy*[2]

Qualitative Sampling of the Building Envelope for Water Leakage

ABSTRACT: Industry standards for intrusive water leakage evaluations of the building envelope are promulgated by ASTM E 2128, "Standard Guide for Evaluating Water Leakage of Building Walls," which lays out a purposeful step-by-step methodology by which information is accumulated and succeeding sampling locations are identified by a skilled professional; however, some laypersons have criticized building envelope survey findings that are not derived statistically from random sampling. This paper notes that there are relatively few building envelope investigations for which statistical random sampling, in and of itself, is a legitimate or practical methodology for achieving a comprehensive understanding of the sources and mechanisms of water leakage and, therefore, the use of quantitative (i.e., statistical) survey protocols to evaluate the validity of purposeful qualitative sampling of the building envelope is not appropriate. Further, this paper demonstrates that a building envelope evaluation that has been carried out in conformance with ASTM E 2128 satisfies current rules of evidence that require an expert's sampling methodology and analysis to be based upon "scientifically valid" principles. In addition, this paper identifies fundamental assumptions that guide the initial steps of most water leakage evaluations, proposes basic categories for prioritizing the sampling, and discusses the potentially complementary roles of limited surveys by plaintiff and defense experts.

KEYWORDS: qualitative sampling, qualitative survey, purposive sampling, inductive analysis, random sampling, building envelope, ASTM E 2128, substantive significance, rules of evidence

Introduction

A common analytical error by layperson reviewers of reported results of forensic evaluation of building envelope construction or performance is the assumption that the only accepted or valid survey technique is blind simple random sampling (SRS), for which all units of the target population have the same probability of being randomly selected for sampling. To better evaluate this assumption, it is helpful to list several key elements of the basic SRS protocol:

a. Sampling locations are predetermined randomly; the sampler has no discretion in the initial selection process.
b. Every member of the "population" being sampled has the same chance to be selected.
c. The entire sampling process is carried out in rigid conformance with the predetermined protocol; i.e., the findings from any particular sample cannot be used in any manner to shape the course of the continued sampling [1].

Upon only brief review, the experienced building envelope professional can identify potentially critical hindrances to truly random statistical sampling of buildings, including:

a. Aesthetic or logistical constraints imposed by the building owner or occupants;
b. Access constraints created by the building's location or configuration;
c. Legal limitations on access; and
d. The high costs imposed by the inflexible sampling design, which necessitates additional rounds of SRS sampling to help clarify the results of the initial sampling.

Another key objection to using an SRS protocol for a general building evaluation is that this simply is not an appropriate method for evaluation of building leakage and causative defects that predominately are found at changes-in-material and changes-in-plane of the building envelope. The anecdotal "1 % Rule" asserts that 99 % of the sources of water infiltration are found at 1 % of the building envelope. While there

Manuscript received September 23, 2006; accepted for publication August 23, 2007; published online September 2007.
[1] Codes & Standards Consultant, Richard Avelar & Associates, 318 Harrison St., Suite 103, Oakland, CA, 94607.
[2] Principal, Trinity ERD, 80 Yesler Way, Suite 200, Seattle, WA, 98104.

certainly is more than a touch of hyperbole to this "rule," it does help make clear that randomly devoting a great majority of the predetermined sampling to areas that are least likely to be damaged or defective is not an effective use of typically limited time and resources. Similarly, the standard requirement of a predetermined SRS design that data obtained during the course of the sampling cannot be used to guide the course of the additional sampling represents a flagrant waste of the building envelope professional's forensic expertise.

It might be argued that some of these objections can be addressed with advanced statistical sampling designs such as "stratified random sampling" and "adaptive cluster sampling":

a. *Stratified random sampling*—The target population is divided into spatial regions and simple random sampling is carried out independently within each region.
b. *Adaptive cluster sampling*—Initial samples are selected randomly. Additional random samples are collected at areas of interest identified in the initial sampling. This process of random sampling iterates until a predetermined threshold of understanding is achieved [2].

However, such alternative approaches tend to sacrifice some level of the true randomness so highly prized by statisticians and yet still cannot be carried within the range of time, access, logistical, budgetary, aesthetic and legal constraints, or a combination thereof, commonly encountered by the building envelope professional.

Despite the predilection of many laypersons to the idealized SRS protocol, any sample survey methodology that is not feasible, practical, or cost effective cannot be considered appropriate or representative of the "scientific method" [3].

Further, random probability sampling and its siblings are not consistent with the building envelope survey standards established by ASTM E 2128, "Standard Guide for Evaluating Water Leakage of Building Walls," which prescribes a purposeful (i.e., nonrandom and typically only partially predetermined) investigation that entails an orderly step-by-step accumulation of information in such a manner that each step enhances and supplements the information gathered in the preceding step [4].

Clearly, the purposive sampling methodology described within ASTM E 2128 is not consistent with the tenets of random *quantitative sampling* (probability sampling), as exemplified by the SRS ideal, and therefore may be subject to criticism that the results of such purposeful surveys do not satisfy current rules of evidence that require an expert's sampling methodology and testimony to be based upon "scientifically valid" principles [5].

This paper addresses this criticism by demonstrating that the survey protocol prescribed by ASTM E 2128 is fully consistent with *qualitative sampling* methodology validated within the social sciences.

(Note: while this paper focuses on qualitative surveys of exterior walls, the discussion and review provided below also are applicable in general to leakage surveys of the remaining major components of a structure's building envelope—i.e., roof covering systems and below grade waterproofing.)

The Fundamentals of Qualitative Sampling

Seminal standards in the field of qualitative inquiry review the different logics that undergird the quantitative and qualitative sampling methodologies. While quantitative methods generally require a large randomly selected sample set, qualitative inquiry typically focuses in depth on relatively small samples, even single cases ($N=1$), selected purposefully. Not only do the sampling techniques differ, but the very logic of each approach is unique because their purpose is different:

a. The logic and power of random sampling derive from statistical probability theory—a random and statistically representative sample set controls for selection bias and permits confident generalization from the sampling to a larger population.
b. In contrast, what would be "bias" in statistical sampling, and therefore a weakness, becomes intended focus in qualitative sampling, and therefore a strength. The logic and power of purposeful sampling lie in selecting *information-rich* samples for step-by-step evaluation of issues of central importance to the purpose of the inquiry. For the experienced researcher, studying information-rich samples yields broad insights and in-depth understanding of the existing population rather than empirical generalizations derived from probability theory [6].

A cardinal principle of qualitative sampling (*aka* inductive analysis) is that resulting causal findings and theoretical statements clearly must be emergent from and grounded in purposive (or judgmental) field observation. In other words, the expert's step-by-step process of qualitative analysis builds toward general patterns that emerge from a series of purposefully collected datasets, instead of being derived statistically from a dataset that has been randomly generated to prove or disprove a predetermined hypothesis [7,8].

While the goal of quantitative sampling is to evaluate levels of *statistical significance*, the methodology of qualitative inquiry is to produce findings that have *substantive significance*, which refers to the strength and importance of a meaningful relationship [9].

In contrast with the meaningful correlations denoted by substantive significance, a finding of statistical significance does not necessarily indicate a relationship that is important or meaningful to the goal(s) or finding(s) of the investigation; it simply means that enough samples were examined to rule out (within a statistically calculable "confidence interval" for a given "confidence level") the effects of chance [10].

The qualitative analyst presents an argument for substantive significance in the findings of a summary report, but subsequent reviewers of this expert's analysis also will make their own value judgments about significance. In determining substantive significance, both the analyst and the reviewers must address these kinds of questions:

a. How solid, coherent, and consistent is the qualitative evidence in support of the expert's findings?
b. To what extent and in what ways do the findings further a deep understanding of the observed conditions?
c. How well do the findings define and correlate causal relationships in a manner that maximizes understanding of the various processes and phenomena of interest that are occurring within the population?
d. To what extent are the expert's findings consistent with knowledge derived from other sources? [11,12].

The sampling design for qualitative surveys typically is fluid, capitalizing on early learning to guide subsequent direction. Qualitative researchers begin by identifying information-rich data sources that can be expected to maximize understanding of the observable conditions. As the survey progresses, new conditions and sampling questions emerge that may confirm, enrich, modify, or challenge the researcher's understanding of the observed phenomena. Qualitative sampling represents an inductive process in which the researcher searches for patterns and builds abstractions, concepts, hypotheses, and theories from emerging details [13].

Unlike probability sampling, there are no firm criteria for determining sample size in qualitative surveys. The appropriate sample size is determined by the quality of the observed data as it relates to the goal of the survey. If the key purpose is to maximize information, then the survey may be terminated when no new information is forthcoming from the additional sampling; thus, *data saturation* is the primary criterion. In other words, for some qualitative surveys, the guiding principle is sampling to the point of redundancy, where no new information of substantive significance is obtained [14,15].

However, sample size for qualitative surveys also may depend simply on what you want to know, the purpose of the inquiry, what will be useful and will have credibility, and what can be done with available time and resources [16].

To accommodate varying goals, conditions, and limitations, social scientists can choose from a wide range of accepted methodologies for qualitative sampling. The underlying principle that is common to these strategies is selecting information-rich samples from which a great deal can be learned about matters of importance [17].

Among the qualitative survey methodologies of potential interest to building envelope leakage investigators is the highly respected "grounded theory," in which data, patterns, and categories, or a combination thereof, emerge through inductive sampling and then are interpreted and affirmed via deductive reasoning [18].

The core of the grounded theory methodology consists of the interplay between making inductions (deriving concepts, their properties, and dimensions from data) and making deductions (hypothesizing about the relationships between concepts) that are empirically "grounded" within the inductive survey process [19].

In contrast, the "analytic induction" methodology begins deductively by formulating a hypothesis that is evaluated by in-depth qualitative examination of a particular case. If the facts of the case are found to support the hypothesis, then additional cases are studied in a quest for generalizations. If not, then the original hypothesis is revised and the investigation continues. The practitioners of analytic induction emphasize close examination of negative cases in order to fine tune hypotheses and conclusions [20].

Other qualitative analysts are proponents of a methodology of "logical analysis" (*aka* "abduction") in which the inductive and deductive models are combined to work backward from consequence (effect) to cause. The investigative process used by famed fictional detective Sherlock Holmes is cited as an example of "abductive analysis" [21–23].

A lesser used methodology for qualitative surveys is an "advocacy-adversary" model in which opposing teams work with the same set of data but each team organizes and interprets this data to support opposite conclusions. The strength of the advocacy-adversary model is that it highlights the importance of engaging in systematic analysis of alternative and rival conclusions. The weakness of this approach is that it can promote a biased advocacy to the detriment of a substantive understanding of the issues [24].

For their findings to be considered substantive, all qualitative researchers, including building envelope professionals, must avoid any biased advocacy that hides, distorts, or selectively interprets the collected data. The mantra of the qualitative researcher must be: "*It is what it is*." In other words, the data tell the story, even if this story differs from the desired or expected findings.

Of the many sampling designs used to implement the qualitative survey methodologies outlined above, the following will be familiar to experienced building envelope professionals:

a. *Intensity sampling*—Select information-rich cases that manifest the conditions intensely, but not extremely. (For example, when identifying a sampling location to evaluate potential sources of water leakage, the experienced professional will avoid a location at which a primary source of leakage likely has resulted in damage so severe that it masks evidence of additional sources of leakage.)
b. *Opportunistic* or *emergent sampling*—Immediately follow new leads exposed during the sampling. The qualitative researcher must have the experience and flexibility to be able to take advantage of the unexpected.
c. *Maximum variation sampling*—Purposely picking samples that likely will exhibit a wide range of variation of expected conditions of interest in order to identify key common patterns that "cut through the noise of variation" and to allow documentation of unique or diverse variations that have emerged during adaptation to these differing conditions.
d. *Confirming* and *disconfirming sampling*—Elaborating upon initial analyses by seeking exceptions and testing variations.
e. *Purposeful random sampling*—Taking purposeful random samples in order to reduce suspicion about the integrity of particular findings or of the overall investigation. A qualitative sampling strategy does not necessarily eliminate any possibility for random selection of samples. The purpose of such qualitative random sampling is credibility, not the generalized statistical representativeness derived from probability sampling [25].

In all such qualitative surveys, the quality and rigor of the resultant data and substantive findings is highly dependent on the skill, training, education, competence, and integrity of the investigator. In short, the substantive value of the qualitative inquiry depends on three distinct but related elements:

a. Rigorous methods for doing fieldwork and collecting high-quality data.
b. Ongoing, thorough, and systematic analyses of these data.
c. The credibility of the researcher [26].

All surveys, whether quantitative or qualitative, require the investigator to have training, integrity and credibility; however, for qualitative surveys the abilities and qualities of the evaluator have increased importance because the processes of data collection and analysis do not rest upon a rigid statistically-derived foundation. While statistical analysis follows formulas and rules, the core of qualitative analysis is composed of the insights, knowledge, and conceptual capabilities of the experienced analyst [27,28].

Note that qualitative and quantitative sampling methodologies constitute alternative, but not mutually exclusive, strategies for investigative research. The experienced, pragmatic investigator practices a situational responsiveness which recognizes that differing methods and techniques are appropriate for different circumstances [29].

As the final step in the qualitative survey process—after describing and interpreting the major patterns, themes, and linkages that have emerged from the analysis—the expert investigator must, as a matter of intellectual integrity, look for data that support alternative themes and explanations for the observed conditions. Failure to find strong supporting evidence for opposing theories increases confidence in the researcher's original analysis. Closely related to such testing of rival explanations is the search for negative cases; comprehensive understanding of the perceived qualitative patterns is increased by considering the instances and cases that do not fit the pattern [30].

ASTM E 2128's Methodology for Qualitative Investigation of the Building Envelope

It is instructive to compare the above discussion of the fundamentals of qualitative sampling with the investigative methodology prescribed in ASTM E 2128. Section 11 of this standard includes, in part, the following protocol for carrying out a qualitative survey of the building envelope:

1. An evaluation is conducted in response to a problem situation and a nonperforming wall, and may involve several techniques and procedures specifically adapted and applied in a systematic manner to diagnose a specific problem.
2. The information systematically accumulated in a leakage evaluation is analyzed as it is acquired. The new information may motivate a change in approach or focus for subsequent steps in the evaluation process.
3. The evaluator is expected to establish a cause and effect relationship between wall characteristics and observed leakage. This requires an appropriate selection of activities and a logical analysis and interpretation of the acquired information.
4. The conclusions and findings from an evaluation must be rationally based on the activities and procedures undertaken and the information acquired, if they are to be considered legitimate and substantiated.
5. The record should be sufficiently complete so that any interested party can duplicate the evaluation program and acquire similar information. Notes on the analysis and interpretation of the acquired information should be clear and complete enough to be understood by any other building professional skilled in leakage evaluation.

ASTM E 2128 clearly is prescribing a purposeful qualitative inquiry that is fully consistent with the qualitative sampling methodologies discussed above. The goal of the skilled building professional is to produce findings of substantive significance that identify cause and effect relationships between building envelope characteristics and observed leakage and resulting damage. To this end, the building professional must provide a record of the investigation and analysis that is sufficiently complete to enable another professional to duplicate the intertwined processes of observation, sampling, and analysis, including onsite qualitative testing (e.g., water testing to recreate or "chase" leaks), where appropriate.

ASTM E 2128 recognizes that such qualitative testing can be an integral part of the evaluation process and should be thought of as a means to verify and extend hypotheses arrived at during both the document review and inspection phases of the building envelope survey using controlled and reproducible procedures. These qualitative tests may be supplemented with the results of quantitative testing (onsite and at a laboratory, or both) of material properties and their weather-resistive performance.

ASTM E 2128 also accepts the many constraints (e.g., time, access, logistical, budgetary, aesthetic and legal limitations or a combination thereof) that can be placed upon the building envelope investigation by noting that any conclusions or expectations about the substantive significance of the findings must be proportional to the defined scope of work and the effort and resources applied to this task.

In short, ASTM E 2128 presents an excellent and well-delineated overview of a fundamental "evaluation program" used (in full or in part) by numerous first-rate building envelope professionals across North America. Further, as noted in this seminal standard, reviewers of a qualitative evaluation of the building envelope should recognize that the overall process encompasses broader objectives than realistically can be accommodated by the pass/fail format afforded by statistical analysis.

For the well-trained building envelope professional, general conformance with the survey protocols published within ASTM E 2128 constitutes a form of analytical rigor comparable to the validated tenets of qualitative analysis practiced within the social science fields. If this survey process and its findings are reviewed during the course of litigation, the rules of evidence clearly have been satisfied when a building envelope expert's sampling methodology and analysis are founded upon ASTM E 2128.

Key Standards and Assumptions for Qualitative Evaluation of the Building Envelope

The authors recommend that to carry out a substantive water leakage evaluation of the building envelope, the skilled professional must understand the basics of building envelope design and how the various exterior wall systems and materials should be installed and integrated to ensure proper long-term weather-resistive service. To these ends, it is important to have good command of knowledge of the following ASTM standards:

a. ASTM E 241, "Standard Guide for Limiting Water-Induced Damage to Buildings;"
b. ASTM E 2112, "Standard Practice for Installation of Exterior Windows, Doors and Skylights;" and
c. ASTM E 2266, "Standard Guide for Design and Construction of Low-Rise Frame Building Wall Systems to Resist Water Intrusion."

In addition, the qualitative water leakage survey often will be founded upon the following fundamental assumptions that may seem obvious and well-validated to building envelope professionals, but which should be identified for layperson reviewers:

a. Typically, the most severe conditions of rainwater damage, if any, are encountered at the weather-exposed elevation(s). Conversely, the greater the degree of weather protection, the greater the likelihood that an elevation is less severely damaged than the weather-exposed elevations.
b. Similarly, partial wall areas that are weather-protected (e.g., by eaves, overhangs, or adjacent structures) are less likely to be damaged by rainwater.
c The transitions between the work carried out by different trades are the most common locations of water infiltration.
d. Building envelope leakage occurs predominately at changes-in-plane and changes-in-material, such as window and door perimeters and wall transitions to decks and balconies.

The building envelope survey also often commences with a preliminary assumption that the project's design, detailing, and quality control will be found to have been generally consistent at all elevations (and all buildings, where applicable). This assumption must be closely evaluated during the course of the sampling. If a significant level of inconsistent construction design or workmanship is exposed, then the survey must be redesigned (and typically expanded) to properly encompass and evaluate the emergent data.

Where not contradicted by the data that are collected from the interviews, site observations, and document review carried out prior to physical sampling, or a combination thereof, the above assumptions commonly are used by building envelope professionals to identify satisfactory information-rich location(s) for commencing the invasive investigation of the structure.

Categorization Strategy for Identifying Appropriate Sampling Locations

In addition, the authors recommend that for purposes of building envelope survey design for typical structures, three categories of sampling locations can be broadly generalized:

1. Category 1 sampling locations typically occur at changes-in-material and changes-in-plane, such as fenestration and deck-to-wall transitions, that can be expected to be highly information-rich due to the number of waterproofing, flashing, sealing, and water-resistive barrier transitions, or a combination thereof, by various trades that can be exposed in sample areas of relative limited size. Other potentially information-rich Category 1 sampling locations that may be identified during preliminary visual surveys include areas of observable structural deformation or significantly deteriorated cladding or unique characteristics of the building's design or exposure. Generally, building envelope sampling should commence at Category 1 locations, usually at the weather-exposed elevations.
2. Category 2 sampling locations, such as vent penetrations and handrail connections, are transitions that likely are not as information-rich as the Category 1 locations, but which can provide key supplemental information regarding consistency of design, details, and quality control, the potential extent/severity of any deterioration, and the quality of the work carried out by some of the lesser trades.
3. Category 3 locations are those that are sampled to address atypical construction or design issues or specific questions or issues of concern that may arise during the course of the sampling. Category 3 sampling often is carried out in the later phases of the survey process, and can serve various survey closeout purposes, including localized sampling intended to test alternative conclusions raised by others and purposeful random sampling for estimating repair quantities and costs or to increase the perceived credibility of the qualitative analysis.

To achieve substantive understanding of conditions at typical projects, varying levels of Category 1 sampling may be required at all four elevations, supplemented by Category 2 and Category 3 sampling at one or more elevations as necessary to reach a level of data saturation appropriate to the scope of commission received from the client.

The (Potentially) Complementary Roles of Plaintiff and Defense Surveys

It is not uncommon during construction defects litigation for the plaintiff(s) and defendant(s) to commission a series of limited qualitative surveys. Often, a preliminary investigation is carried out by the plaintiff's expert, typically under the close scrutiny of experts hired by the defendants. At a later period, after review of the findings reported by the plaintiff's expert, one or more of the defense experts is commissioned to carry out supplemental surveys (typically under the close scrutiny of the plaintiff's expert) that often evaluate only those issues raised in the plaintiff's report. Even though the scope of each of these commissions may be limited, such limitations are not sufficient, per se, to justify an argument that the reported findings do not have substantive value and merit.

This series of complementary qualitative sampling by opposing parties is consistent with the methodologies of "confirming and disconfirming sampling" and the "advocacy-adversary" model discussed above. Further, while each of these experts may have a differing interpretation of the data exposed during the jointly observed surveys, such differences of opinions also are not inconsistent with qualitative survey standards and often may assist opposing experts to reach (or revise) their findings of substantive significance. However, to achieve substantive findings from this series of complementary qualitative surveys, a cardinal principle, per ASTM E 2128, is that all parties must have equal access to data records that are sufficiently clear and complete so that any interested party skilled in leakage evaluation can duplicate the evaluation process used by any other party.

Similarly, all opposing parties, in principle, should be allowed an opportunity to closely witness the surveys carried out by others. In practice, there can be legitimate and compelling reasons, such as emergency repairs, for carrying out some surveys without providing appropriate notice to other parties. In these cases, per ASTM E 860[3], it is incumbent upon the investigator to preserve documentation supporting the compelling justification for such non-noticed action.

As previously discussed, for their findings to be considered substantive, building envelope professionals must avoid any degree of biased advocacy that hides, distorts, or selectively interprets the collected data. An absolute cornerstone of the qualitative survey process is the credibility of the researcher. The trustworthiness of the qualitative analysis corresponds directly to the trustworthiness and demonstrated competence of the investigator [31].

No scientific rigor should be attributed to findings presented by a building envelope professional who previously has demonstrated a willingness to promote distorted or false data.

The potentially critical weakness of this advocacy-adversary model is exemplified by those building envelope professionals whose narrow focus in litigation cases simply is to mine the database only for information that advocates their predetermined position. Similarly, the potential strength of the model is demonstrated when an experienced "special master" or arbitrator, faced with dueling advocates willing to risk their professional integrity to support biased findings, remains able to extract substantive conclusions from close analysis of the rival presentations.

Summary Discussion

The rules of evidence require the forensic building envelope professional's survey methodology and testimony to be based upon recognized, scientifically valid principles. Where its survey methodology is properly designed and implemented, qualitative analysis, such as the investigative protocols prescribed by the authors of ASTM E 2128, constitutes a form of rigor that is parallel to the rigor of statistical designs aimed at eliminating rival explanations [32].

In the end, the core of the scientific method for all quantitative and qualitative surveys consists of a puzzle-solving strategy or methodology for analysis (or elimination) of rival explanations or hypotheses. This strategy may start its puzzle solving with hypothesis (i.e., quantitative analysis) or it may start with

[3]ASTM E 860, "Standard Practice for Examining and Testing Items That Are or May Become Involved in Litigation."

evidence (i.e., qualitative analysis). The quantitative survey begins by forming a hypothesis that can be evaluated statistically upon later collection of evidence, while the qualitative survey starts the puzzle-solving process by collecting evidence from which substantive explanations will emerge. In both cases, the core of the scientific method is represented by the strategy of analysis of plausible rival hypotheses [33,34].

Permeating throughout the building envelope survey strategy is the critical goal of tying together cause(s) and effect(s). Within the limits of the investigator's commission: (a) the consequences of leakage are established; (b) the severity, consistency and distribution of these consequences are determined; and (c) leakage pathways from (d) construction defects and other types of building envelope failures are identified. This investigative process commonly is both inductive and deductive and should be carried out with methodological competence, intellectual rigor, and professional integrity. The pragmatic investigator will implement a range of qualitative and quantitative measures that best evidences credibility when reviewed by the target audience.

Industry standards for intrusive building envelope surveys are promulgated by ASTM E 2128, which lays out a purposeful step-by-step methodology by which information is accumulated and succeeding sampling locations are identified. The expert's findings then must be reported in a professional manner that describes the conditions under which the evaluation was conducted, the methodology used, the observations and measurements made, and the findings and conclusions. This record should be sufficiently complete so that any interested party skilled in building leakage evaluation can duplicate the evaluation program and acquire similar information. The methodology outlined in ASTM E 2128 is fully consistent with the practical experience of numerous building envelope professionals and with qualitative survey protocols established and validated within the social sciences. A cardinal principle of all such qualitative surveys is that their causal conclusions must be emergent from a purposeful series of field observations by a skilled investigator that capitalize on early learning to guide subsequent direction toward findings that have substantive significance—i.e., a strong meaningful relationship.

In contrast, the methodology of ASTM E 2128 is not consistent with probability surveys, such as simple random sampling and its siblings (e.g., stratified random sampling and adaptive cluster sampling). In many cases, probability survey results often must be stratified and analyzed with a weighted distribution system that, in principle, should be developed prior to any building inspection; however, for all weighted surveys (and, in the authors' opinion, building envelope surveys in particular) the actual representativeness of any predetermined weighting system is far from inevitable [35].

In short, there are relatively few building envelope investigations for which random probability sampling, in and of itself, is a legitimate or practical methodology for achieving a comprehensive understanding of the sources and mechanisms of water leakage. Instead, a forensic building envelope professional's survey methodology and testimony typically should be based upon recognized, scientifically valid principles of qualitative analysis, as exemplified by ASTM E 2128.

References

[1] Freedman, D., Pisani, R., Purves, R., and Adhikari, A., *Statistics*, 2nd ed., W. W. Norton & Co, New York, 1991.

[2] Gilbert, R. O. and Pulsipher, B. A., "Role of Sampling Designs in Obtaining Representative Data," *Environmental Forensics*, Taylor & Francis Group, Philadelphia, PA, Vol. 6, 2005, pp. 27–33.

[3] Barnett, V., *Sample Survey Principles & Methods*, 3rd ed., Oxford University Press, New York, 2002, Chap. 1.

[4] ASTM Standard E 2128, "Standard Guide for Evaluating Water Leakage of Building Walls," *Annual Book of ASTM Standards*, ASTM International, West Conshohocken, PA, 2001.

[5] *Daubert v. Merrell Dow Pharmaceuticals, Inc.*, U. S. Supreme Court, 509 U.S. 579, No. 92–102, Washington, D.C., 1993, and *Kumho Tire Company v. Patrick Carmichael*, U.S. Supreme Court, 526, U.S. 137, No. 97–1709, Washington, D.C., 1999.

[6] Patton, M. Q., *Qualitative Research & Evaluation Methods*, 3rd ed., Sage Publications, Thousand Oaks, CA, 2002, p. 46.

[7] Patton, M. Q., *How to Use Qualitative Methods in Evaluation*, Sage Publications, Newbury Park, CA, 1987.

[8] Strauss, A. and Corbin, C., *Basics of Qualitative Research*, Sage Publications, Newbury Park, CA,

1990.

[9] Vogt, W. P., *Dictionary of Statistics and Methodology*, Sage Publications, Newbury Park, CA, 1993.

[10] Patton, M. Q., *Qualitative Research & Evaluation Methods*, 3rd ed., Sage Publications, Thousand Oaks, CA, 2002.

[11] Patton, M. Q., *Qualitative Research & Evaluation Methods*, 3rd ed., Sage Publications, Thousand Oaks, CA, 2002, p. 467.

[12] Strauss, A. and Corbin, C., *Basics of Qualitative Research*, Sage Publications, Newbury Park, CA, 1990.

[13] Creswell, J. W., *Research Designs—Qualitative & Quantitative Approach*, Sage Publications, Thousand Oaks, CA, 1994, p. 145.

[14] Lincoln, Y. S. and Guba, E. G., *Naturalistic Inquiry*, Sage Publications, Newbury Park, CA, 1985, p. 202.

[15] Lincoln, Y. S. and Guba, E. G., *Naturalistic Inquiry*, Sage Publications, Newbury Park, CA, 1985, p. 188.

[16] Patton, M. Q., *Qualitative Research & Evaluation Methods*, 3rd ed., Sage Publications, Thousand Oaks, CA, 2002, p. 244.

[17] Patton, M. Q., *Qualitative Research & Evaluation Methods*, 3rd ed., Sage Publications, Thousand Oaks, CA, 2002, p. 242.

[18] Patton, M. Q., *Qualitative Research & Evaluation Methods*, 3rd ed., Sage Publications, Thousand Oaks, CA, 2002, p. 454.

[19] Strauss, A. and Corbin, C., *Basics of Qualitative Research*, Sage Publications, Newbury Park, CA, 1990, p. 23.

[20] Patton, M. Q., *Qualitative Research & Evaluation Methods*, 3rd ed., Sage Publications, Thousand Oaks, CA, 2002, p. 94, and 2002, p. 554.

[21] Patton, M. Q., *Qualitative Research & Evaluation Methods*, 3rd ed., Sage Publications, Thousand Oaks, CA, 2002, p. 470.

[22] Denzin, N. K., *The Research Act: A Theoretical Introduction to Sociological Methods*, McGraw-Hill, New York, NY, 1978, p. 110.

[23] Sanders, W., *The Sociologist as Detective*, Praeger Publishers, New York, NY, 1976.

[24] Patton, M. Q., *Qualitative Research & Evaluation Methods*, 3rd ed., Sage Publications, Thousand Oaks, CA, 2002, p. 554.

[25] Patton, M. Q., *Qualitative Research & Evaluation Methods*, 3rd ed., Sage Publications, Thousand Oaks, CA, 2002, pp. 230–244.

[26] Patton, M. Q., *Qualitative Research & Evaluation Methods*, 3rd ed., Sage Publications, Thousand Oaks, CA, 2002, p. 584.

[27] Patton, M. Q., *Qualitative Research & Evaluation Methods*, 3rd ed., Sage Publications, Thousand Oaks, CA, 2002.

[28] Strauss, A. and Corbin, C., *Basics of Qualitative Research*, Sage Publications, Newbury Park, CA, 1990.

[29] Patton, M. Q., *Qualitative Research & Evaluation Methods*, 3rd ed., Sage Publications, Thousand Oaks, CA, 2002, p. 14 and 2002, pp. 71–72.

[30] Patton, M. Q., *How to Use Qualitative Methods in Evaluation*, Sage Publications, Newbury Park, CA, 1987, p. 160.

[31] Patton, M. Q., *Qualitative Research & Evaluation Methods*, 3rd ed., Sage Publications, Thousand Oaks, CA, 2002, p. 570.

[32] Patton, M. Q., *Qualitative Research & Evaluation Methods*, 3rd ed., Sage Publications, Thousand Oaks, CA, 2002, pp. 553–554.

[33] Campbell, D., foreword to Yin, R. K., *Case Study Research: Design and Methods*, Sage Publications, Beverly Hills, CA, 1989.

[34] Patton, M. Q., *Qualitative Research & Evaluation Methods*, 3rd ed., Sage Publications, Thousand Oaks, CA, 2002.

[35] Barnett, V., *Sample Survey Principles & Methods*, 3rd ed., Oxford University Press, New York, 2002, Chap. 1.

Journal of ASTM International, Vol. 5, No. 3
Paper ID JAI100817
Available online at www.astm.org

Gerald A. Dalrymple, P.E.[1] *and A. Rhett Whitlock, Ph.D.*, P.E.[1]

Applying C1601-06 "Standard Test Method for Field Determination of Water Penetration of Masonry Wall Surfaces" in Practice

ABSTRACT: Water permeance of exterior masonry wall systems drives masonry restoration and repair design decisions from selection of flashing replacement to pointing to application of water repellents. ASTM C1601 was first issued in August 2004 and has not been used extensively in service by professionals to evaluate masonry wall surfaces. ASTM C1601 provides a standard method of measuring water permeance that can be used to evaluate existing wall surfaces prior to specifying repair methods and as a tool to evaluate the overall effectiveness of repair methods so that efficient repair options can be selected. This paper reviews the requirements of the test method, application of the test method and interpretation of test results. Historical test data are presented that illustrate use of the test method to evaluate wall systems and expected range of test results for existing and repaired wall systems that can be used to select appropriate repair procedures.

KEYWORDS: water penetration, leakage, masonry, field testing, wind-driven rain, wind pressure

Introduction

ASTM C1601-06 "Standard Test Method for Field Determination of Water Penetration of Masonry Wall Surfaces" [1] was developed through ASTM Committee C15 and was first adopted by the Society in 2004. The current version of the standard was subsequently revised and reissued in 2006. The field test method outlined in the standard was developed based on the test apparatus and general procedures specified in ASTM E514 "Standard Test Method for Water Penetration and Leakage Through Masonry." ASTM E514 was first published in 1974 and is a laboratory test method performed on laboratory constructed masonry wall panels. Over time, many consultants began adapting the test apparatus and laboratory test procedures from ASTM E514 for field use to assist in the evaluation of masonry wall system performance related to water penetration. Use of the ASTM E514 method in the field grew in popularity, and the results obtained between testing groups were often not comparable due to variations in test procedures. Subcommittee C15.04 on Research started developing the field test method in the 1980s, but work was started and stopped several times. Eventually, momentum in development of the field test standard was gained, and the standard was completed in 2004.

Though the equipment and procedures of ASTM C1601-06 and ASTM E514-06 [2] are similar, the results are not directly comparable. Both test methods can be used on any masonry wall system, i.e., single wythe, multi-wythe, cavity walls, and composite walls. The setup of the test system varies slightly for different wall configurations. Surface penetration, as determined by ASTM C1601, is defined as the amount of water passing through the exterior wall surface as measured by a closed loop water flow and water volume measuring system. This property may not be directly comparable to water penetration and leakage as measured by ASTM E514, which are typically defined as the amount of water traveling completely through a masonry wall system and collected on a flashing system on the side opposite the tested surface. In other words, one test measures water volume that goes into the wall, whereas the other measures water volume that passes through the wall and is expelled from the wall. One might think that what flows in must come out, but this is not necessarily so, as water may flow within the wall and exit from the tested face of the wall.

Some factors that affect the rate of water penetration are the exposure of a wall to the elements and

Manuscript received September 25, 2006; accepted for publication January 17, 2008; published online March 2008.
[1] Principal, Whitlock Dalrymple Poston & Associates, P.C., 10621 Gateway Blvd., Suite 200, Manassas, VA 20110, USA.

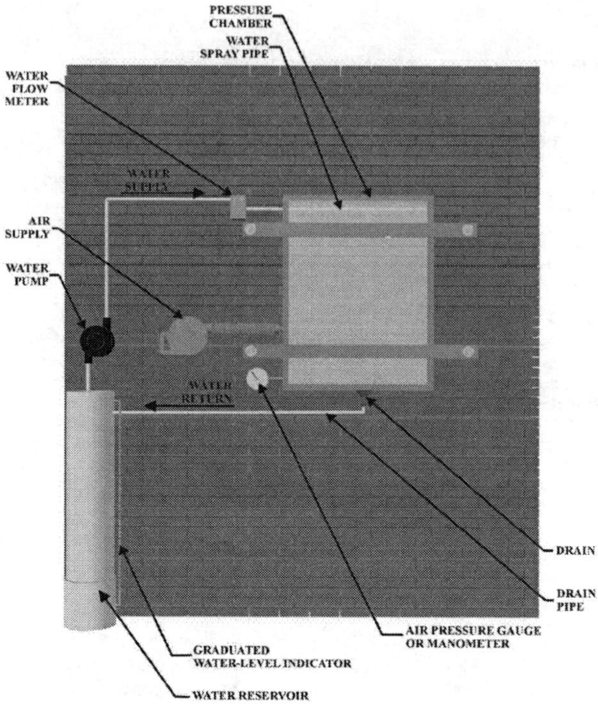

FIG. 1—*ASTM C1601 field test apparatus setup.*

local climatology, the size of masonry units, type of masonry, cracks in masonry units and mortar, solidity of head and bed joints, workmanship, compatibility of masonry units and mortar (properties of mortar and masonry unit characteristics), mortar shrinkage, and any other effects that would disturb intimate bonding between the masonry units and mortar.

ASTM C1601 Test Procedures

The test method uses a chamber mounted to the masonry surface to simultaneously apply a water spray and air pressure to the masonry surface to simulate wind-driven rain. The typical apparatus setup in the field is shown in Fig. 1.

The testing apparatus of ASTM C1601 consists of a chamber which is typically constructed of wood framing with an acrylic (Plexiglas®) or a polycarbonate (Lexan®) facing. The chamber is mechanically anchored to the wall surface, and the perimeter of the chamber is sealed to the wall surface to prevent water escape from the chamber perimeter during testing. The top of the chamber is fitted with a water spray bar connected to a small pump and water flow regulator. Water is pumped from a reservoir into the chamber at a specified application rate. Water flows down the wall surface in a sheet, is collected at the base of the chamber, and returned to the reservoir. As water is applied to the wall surface, a specified positive air pressure is applied into the chamber by means of an air blower or compressor with an air flow regulator. The air pressure inside of the chamber is measured with a manometer or pressure gage.

Water flows in a closed loop between the reservoir and the chamber under positive air pressure within the chamber. The volume of water in the reservoir is measured during the testing, and the loss of water from the reservoir over time is recorded as the water penetration rate through the masonry test surface area.

Standard ASTM C1601 test conditions are:

- Positive Air Pressure: 10 psf (500 Pa)

FIG. 2—*Custom sized chambers fabricated to fit project conditions.*

- Water Application Rate: 3.4 gal/ft^2/h (138 L/m^2/h)
- Preconditioning Duration: 30 min
- Test Duration: 4 h

The standard test conditions represent a rain storm equal to approximately 5-1/2 in. (13.97 cm) of rain per hour accompanied by sustained 62.5 mph (100.6 km/h) wind over a 4 h duration. The standard test conditions are severe, but their use dates back to 1974 when they were incorporated into the ASTM E514 standard laboratory test method, so there is a significant amount of historic data available for masonry tested under conditions similar to these.

The preconditioning requirement was developed to satisfy the normal water absorption of the masonry area being tested. With the basic absorption of the masonry satisfied, water loss during the test represents water passing through the masonry wythe into the wall system.

The size of the test chamber can be varied. The minimum test area is 12 ft^2 (1.08 m^2) with a minimum side dimension of 2 ft (0.6 m). We have built custom chambers in the past to test varying sections of masonry. Figure 2 shows chambers custom built to fit spandrel panels below strip windows. The limiting factors affecting the maximum size of custom built chambers are typically the size of the water reservoir to supply the test area and pump capacity required to maintain the specified water application rate over the test area. The intent of the test method is that a sheet of water be developed and maintained over the wall surface during testing. Thus, odd chamber sizes or geometries that do not permit development of a sheet of water over the wall surface do not meet the intent of the test method and would likely produce inaccurate results.

The test method also permits use of alternate test conditions. Test conditions can be specified based on local climate conditions or other useful testing parameters. We have typically tested wall sections at the standard water flow rate at air pressures of 10 psf (500 Pa), 5 psf (250 Pa), and no air pressure to represent different levels of wind-driven rain. The 10 and 5 psf (500 and 250 Pa) air pressures represent sustained wind speeds of 62.5 mph and 44 mph (100.6 and 70.8 km/h), respectively.

The test method does not include performance criteria or performance ratings. Performance requirements must be specified by the user, or the method can be used to compare performance between systems or products or to evaluate effectiveness of different repair procedures on specific projects.

Field Test Statistics and Trends

Table 1 contains the results of 270 field tests performed in-house by our firm. All of the tests were performed on brick masonry walls except for Projects J, V, AA, DD, EE, and FF, which were natural stone masonry, and Project X which was Portland cement plaster (stucco). Many of the tests were performed before the ASTM C1601 test standard was adopted, but the testing was performed in general conformance with the adopted test method.

Most of the locations were tested using air pressures of 0, 5, and 10 psf (0, 250, and 500 Pa).

TABLE 1—*Water Penetration Test Results.*

		Permeance Rates, L/hr						
		Before Surface Repairs/Treatment			After Surface Repairs/Treatment			
Project	Test location	0 psf	5 psf	10 psf	0 psf	5 psf	10 psf	Repair Remarks
A	1	22.7	59.1	73.5				
brick	2	13.2	31.3	42				
masonry	Avg.	17.95	45.2	57.75				
	Std. Dev.	6.72	19.66	22.27				
	C.O.V.	37.42 %	43.49 %	38.57 %				
B[a]	1	5.11		10.34	1.06		5.29	Water repellent
brick	2	4.47		10.66				
masonry	3	68.01		72.30				
	Avg.	25.86		31.10				
	Std. Dev.	36.50		35.68				
	C.O.V.	141.13 %		114.73 %				
C	1			28.40			4.00	Water repellent only
brick	2			45.10			3.20	Partial pointing + water repellent
masonry	3			27.90			0.81	Full pointing + water repellent
	Avg.			33.80			2.67	
	Std. Dev.			9.79			1.66	
	C.O.V.			28.96 %			62.16 %	
D[b]	1	0.18	0.90	1.88	0.24	0.24	0.37	Water repellent
brick	2	0.60	1.20	1.01	0.24	0.36	0.26	Water repellent
masonry	3	0.78	1.62	2.01			1.72	Water repellent
	Avg.	0.52	1.24	1.63	0.24	0.30	0.79	
	Std. Dev.	0.31	0.36	0.54	0.00	0.08	0.81	
	C.O.V.	59.21 %	29.17 %	33.29 %	0.00	28.28 %	103.30 %	
E	1			7.50 %			1.90	Partial pointing + water repellent
brick	2			6.20			2.40	Full pointing + water repellent
masonry	3			9.20			4.30	Water repellent only
	Avg.			7.63			2.87	
	Std. Dev.			1.50			1.27	
	C.O.V.			19.71 %			44.17 %	
F	1	2.84	4.31	5.22				
brick	2	1.93		7.87				
masonry	3					0.15	0.39	Tuckpointing only
	4					0.15	0.46	Tuckpointing only
	Avg.	2.39		6.55		0.15	0.43	
	Std. Dev.	0.64		1.87		0.00	0.05	
	C.O.V.	26.98 %		28.63 %		0.00 %	10.65 %	
G	1			8.58			5.40	Water repellent
brick	2			19.73			12.37	Water repellent
masonry	3			3.23			1.38	Water repellent
	Avg.			10.51			6.38	
	Std. Dev.			8.42			5.56	
	C.O.V.			80.07 %			87.11 %	
H	1			24.94			12.97	Water repellent
brick							5.76	Water repellent
masonry	Avg.						9.37	
	Std. Dev.						5.10	
	C.O.V.						54.44 %	

TABLE 1— *(Continued.)*

		Permeance Rates, L/hr						
		Before Surface Repairs/Treatment			After Surface Repairs/Treatment			
Project	Test location	0 psf	5 psf	10 psf	0 psf	5 psf	10 psf	Repair Remarks
I[b]	1	0.44	0.65	0.84				
brick	2		1.81	2.88				
masonry	3	0.35		0.55				
	4	0.37	0.61	0.94				
	Avg.	0.39	1.02	1.30				
	Std. Dev.	0.05	0.68	1.06				
	C.O.V.	12.22 %	66.60 %	81.73 %				
J	1	30.57	39.30	56.21				
stone	2	33.48	43.63	49.60	2.79	7.03	7.06	Tuck-pointed only
masonry	3	25.55	60.80	70.20	0.77	2.17	2.03	Tuck-pointed only
	Avg.	29.87	47.91	58.67	1.78	4.6	4.55	
	Std. Dev.	4.01	11.37	10.52	1.43	3.44	3.56	
	C.O.V.	13.43 %	23.73 %	17.93 %	80.24 %	74.71 %	78.26%	
K	1		9.45	14.09				
brick								
masonry								
L	1	0.00	2.27	4.54				
brick	2			23.85				
masonry	3	0.53	1.93	2.65				
	4	0.30	1.02	3.40				
	Avg.	0.42	1.74	8.61				
	Std. Dev.	0.16	0.65	10.19				
	C.O.V.	39.19 %	37.14 %	118.35 %				
M	1		13.20					
brick	3		14.70					
masonry	4		20.00					
	5		12.10					
	6		3.00	9.40				
	7	8.70	17.00	22.30				
	8	14.70	27.60	36.30				
	9a	13.60	23.80	32.90				
	9b	13.60	23.10					
	Avg.	12.65	17.17	25.23				
	Std. Dev.	2.68	7.45	12.12				
	C.O.V.	21.22 %	43.41 %	48.04 %				
N	1	11.20	22.33	22.33				
brick	2	4.31	7.53	10.90				
masonry	3	8.10	15.97	17.60				
	4	5.53	15.75	13.17				
	Avg.	7.29	15.40	16.00				
	Std. Dev.	3.05	6.07	5.05				
	C.O.V.	41.88 %	39.41 %	31.59 %				
O	1	17.70	22.90	36.30				
brick	2	14.20	28.60	39.40				
masonry	3	8.50	16.20	26.60				
	Avg.	13.47	22.57	34.10				
	Std. Dev.	4.64	6.21	6.68				
	C.O.V.	34.48 %	27.50 %	19.58 %				

TABLE 1— *(Continued.)*

		Permeance Rates, L/hr						
		Before Surface Repairs/Treatment			After Surface Repairs/Treatment			
Project	Test location	0 psf	5 psf	10 psf	0 psf	5 psf	10 psf	Repair Remarks
P	1	3.09	4.94	7.41				
brick								
masonry								
Q	1	14.76	25.93	37.98				
Brick								
masonry								
R	1			10.00				
brick	2			3.50				
masonry	3			5.10				
	4			7.20				
	Avg.			6.45				
	Std. Dev.			2.81				
	C.O.V.			43.57 %				
S	1			41.10				
brick	2			25.10				
masonry	3			36.00				
	Avg.			34.07				
	Std. Dev.			8.17				
	C.O.V.			23.99 %				
T	1			72.90				
brick	2			82.50				
masonry	3			54.70				
	Avg.			70.03				
	Std. Dev.			14.12				
	C.O.V.			20.16 %				
U	1	2.16	6.72	8.52				
brick	2		0.46	0.23				
masonry	3	1.46	3.16	3.82				
	Avg.	1.81	3.45	4.19				
	Std. Dev.	0.50	3.14	4.16				
	C.O.V.	27.44 %	91.03 %	99.18 %				
V[c]	1			297.86				
stone	2		94.41					
masonry								
W[a,d]	1	10.09	31.56	42.46				
brick	2	30.45	59.55	71.24				
masonry	3	17.33	39.63	50.02				
	4	33.49		100.47				
	Avg.	22.84	43.58	66.05				
	Std. Dev.	11.02	14.41	25.98				
	C.O.V.	48.24 %	33.06 %	39.34 %				
X[e]	1			0.57				
stucco	2			0.47				
	Avg.			0.52				
	Std. Dev.			0.07				
	C.O.V.			0.05 %				

TABLE 1— *(Continued.)*

		Permeance Rates, L/hr						
		Before Surface Repairs/Treatment			After Surface Repairs/Treatment			
Project	Test location	0 psf	5 psf	10 psf	0 psf	5 psf	10 psf	Repair Remarks
Y	1	13.97	36.55	48.00				
brick	2	18.91	34.13	35.86				
masonry	Avg.	16.44	35.34	41.93				
	Std. Dev.	3.49	1.71	8.58				
	C.O.V.	2.38 %	1.16 %	5.84 %				
Z	1	7.86	13.15	15.60				
brick	2	9.19	26.11	44.47				
masonry	Avg.	8.53	19.63	30.04				
	Std. Dev.	0.94	9.16	20.41				
	C.O.V.	0.64 %	6.23 %	13.89 %				
AA	1	4.40	10.70	17.70				
stone	2	3.60	6.70	9.50				
masonry	3	38.50	52.40	64.20				
	Avg.	15.50	23.27	30.47				
	Std. Dev.	19.92	25.31	29.50				
	C.O.V.	13.55 %	17.22 %	20.07 %				
BB	1	7.00	16.20	20.40				
brick	2	6.00	13.10	17.70				
masonry	3	6.80	13.60	19.80				
	4	12.90	30.70	38.40				
	5	6.00	10.70	15.90				
	6	9.30	20.10	25.90				
	Avg.	8.00	17.40	23.02				
	Std. Dev.	2.69	7.26	8.26				
	C.O.V.	1.83 %	4.94 %	5.62 %				
CC[f]	1	7.02	9.72	11.85				
brick	2	5.45	7.19	8.07				
masonry	3	1.77	10.35	35.71				
	4	3.23	8.77	15.80				
	Avg.	4.37	9.01	17.86				
	Std. Dev.	2.33	1.37	12.31				
	C.O.V.	1.58 %	0.94 %	8.38 %				
DD	1	10.60	17.20	25.10				
stone	2	5.00	12.90	15.10				
masonry	Avg.	7.80	15.05	20.10				
	Std. Dev.	3.96	3.04	7.07				
	C.O.V.	2.69 %	2.07 %	4.81 %				
EE stone masonry	1	11.70	32.00	47.20				
FF	1	8.84	12.46	15.12				
stone	2	7.61	15.13	20.22				
masonry	3	8.67	25.89	36.99				
	Avg.	8.37	17.83	24.11				
	Std. Dev.	0.67	7.11	11.44				
	C.O.V.	0.45 %	4.84 %	7.78 %				

TABLE 1— *(Continued.)*

		Permeance Rates, L/hr						
		Before Surface Repairs/Treatment			After Surface Repairs/Treatment			
Project	Test location	0 psf	5 psf	10 psf	0 psf	5 psf	10 psf	Repair Remarks
GG brick masonry	1	39.00	93.20	111.50				
HH	1	1.70	8.60	6.30				
brick	2	1.60	3.20	4.40				
masonry	3	1.40	2.80	5.00				
	Avg.	1.57	4.87	5.23				
	Std. Dev.	0.15	3.24	0.97				
	C.O.V.	0.1 %	2.20 %	0.66 %				
II	1	13.58	25.75	59.92				
brick	2	5.64	28.71	53.88				
masonry	3	5.11	12.68	15.43				
	Avg.	8.11	22.38	43.08				
	Std. Dev.	4.74	8.53	24.13				
	C.O.V.	3.23 %	5.80 %	16.42 %				
JJ	1			12.97			41.08	Water repellent
brick	2			23.52			8.38	Water repellent
masonry	3			29.88			7.56	Water repellent/spot tuck pointing
	Avg.			22.12			19.01	
	Std. Dev.			8.54			19.12	
	C.O.V.			5.81 %			13.01 %	

Test pressures of 5 psf and 10 psf are equivalent to 250 pa and 500 pa.

[a]Proj. B - Test Loc. 3 and Proj. W - Test Loc. 4 contained large crack/cracked joint.

[b]Water repellent used in original construction.

[c]Test 1 - Due to high permeabilty, water flow into chamber was increased from 0.68 gal/min to 1.75 gal/min (2.57 L/min to 6.62 L/min). Test 2 - Maximum attainable pressure was 2.7 psf (130 Pa).

[d]Project had extruded mortar joints.

[e]Stucco.

[f]Heavily damaged area.

Observed Penetration Rates

For all tests, the rates of water penetration for masonry walls ranged from less than 0.2 L/h to almost 300 L/h. Table 2 shows the ranges for various materials tested. The average penetration rates for each project are shown in Table 1.

TABLE 2—*Water penetration rate ranges, L/h*[a].

	0 psf		5 psf (250 Pa)		10 psf (500 Pa)	
Type	Low	High	Low	High	Low	High
Brick	0	68.01	0.46	93.20	0.23	111.50
Stone	3.60	38.50	6.70	94.41	9.50	297.86
Brick repaired	0.24	1.06	0.15	0.36	0.26	12.97
Stone repaired	0.77	2.79	2.17	7.03	2.03	7.06
Stucco	...	...	...	...	0.47	0.57

[a]Results from the test method have historically been reported in mixed units, i.e., penetration rate in L/h and test pressure in psf. The convention has been retained here.

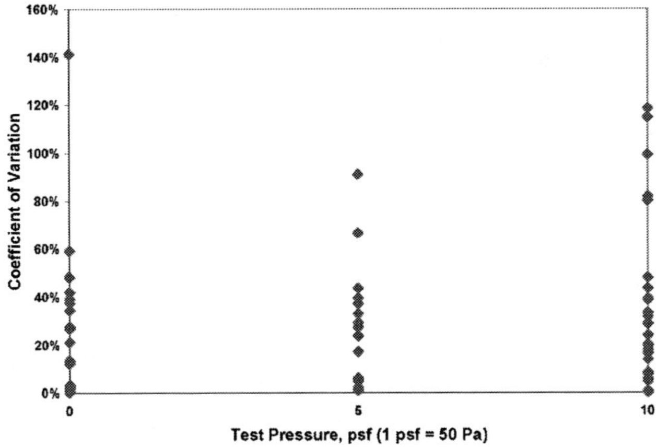

FIG. 3—*Coefficients of variation versus pressure before repairs for all data from Table 1.*

Considering all brick masonry projects together excluding tests over structural cracks, in heavily damaged areas, stone, and a project with extruded mortar joints, the average water penetration rates were 7.54, 15.85, and 21.98 L/h for 0, 5, and 10 psf (0, 250, and 500 Pa), respectively, for unrepaired walls and 0.51, 0.23, and 6.00 L/h for repaired walls.

As can be seen from Table 2, the penetration rate at 0 pressure was on average approximately one-third of the rate at 10 psf (500 Pa).

The stucco tests were performed over an uncracked surface with a few visible small holes ("pinholes"). The results of the stucco tests confirmed that uncracked stucco is relatively water resistant, but based on the authors' experience doing hose and spray rack tests on other projects, cracked stucco leaks profusely.

Variation within a Project

For a normal construction project, the common elements are specific brick, specific mortar, and mason contractor. The quality of each can vary somewhat throughout a project. Environmental conditions, such as temperature, wind, and precipitation, are also subject to variation. Certain properties of each material, workmanship, and the environment can affect water penetration rates.

The coefficients of variation (COV) from each project are plotted in Figs. 3 and 4 for walls tested before and after repairs. As can be seen, the COV is quite large and overall does not tend to significantly change with air pressure. On average, the COV for penetration rate of a building wall was 41.90, 43.45, and 49.30 % for 0, 5, and 10 psf (250 Pa and 500 Pa), respectively, prior to repairs and 40.12, 34.33, and 62.87 % after repairs.

Water Permeability Rating Systems

A rating system for water permeance of masonry wall systems has been a subject of much debate for many years. Several rating systems have been published and are described in the following sections.

BMS 7—Water Permeability of Masonry Walls

The first extensive research program on water permeability of masonry walls was performed by the National Bureau of Standards (NBS), now known as National Institute for Standards and Technology (NIST). The report of the research was entitled "Building Materials and Structures Report BMS 7 - Water Permeability of Masonry Walls" published in 1938 [3]. The testing program included three kinds of brick,

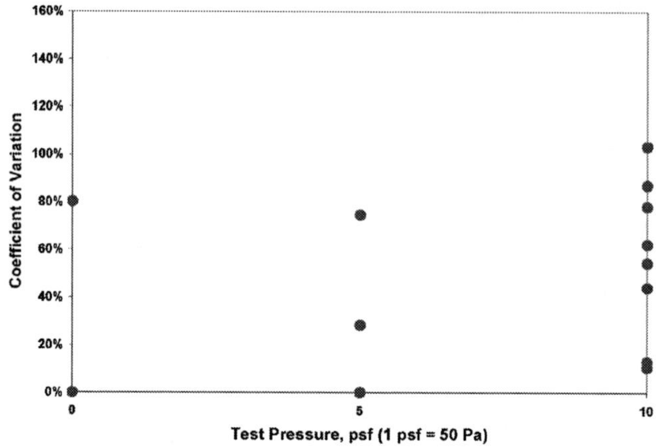

FIG. 4—*Coefficients of variation versus pressure after repairs for all data from Table 1.*

six kinds of structural clay tile, two kinds of concrete masonry, five levels of workmanship, and six mortar types. One hundred thirteen (113) walls were built and tested. The apparatus and procedures were very similar to the current ASTM E514 test method.

In the BMS 7 report, a rating system was developed and was characterized by the report as "arbitrary," as its use was intended for comparison among tests; and the lengthy exposure during testing was "more severe than natural exposure for most building walls." The BMS 7 rating system is shown in Table 3.

The method of water collection was through use of two flashings, one collected water at the very bottom of both wythes and the other collected water that flowed down the interior surface of the inner (back-up wythe). Hence, the sum of both quantities would represent the total amount that flowed through the outer facing.

As can be seen, the rating system was unusual, as there was a very small difference between "Excellent" and "Fair," 0 L/h to 1 L/h and a large difference between "Fair" and "Poor" to "Very Poor," 1 L/h to 16 L/h. It should be noted that the walls tested were 8 in. to 12 in. (20.32 cm to 30.48 cm) thick and were constructed and cured under laboratory conditions.

ASTM E514-74

The first edition of the laboratory test method, E514-74 "Water Penetration and Leakage Through Masonry" [4], contained a rating system that closely resembled that of the BMS 7 report. The test method included 1 day of preconditioning and 3 subsequent days of exposure to test conditions or testing until a rating as defined below was obtained.

The E514 rating system was:

- Class E - No water visible on back of wall above flashing at the end of 1 day. Not more than 25 % of wall area above flashing damp at the end of 3 days. No leaks through wall at the end of 1 day.

TABLE 3—*BMS 7 rating system.*

	Leakage, L/h			
Rating	Outer Wythe	Inner Wythe	Total Leakage, L/h	Inner Surface Dampness, %
E-Excellent	0	0	0	<25
G-Good	0	0	0	<50
F-Fair	<1	<1	<1	>50
P-Poor	<15	<1	<16	NA
VP-Very Poor	>15	>1	>16	NA

A leak was defined by the standard as a flow of water from the flashing at a rate equal to or greater than 0.05 L/h.

- Class G - No water visible on back of wall above flashing at the end of 1 day. Less than 50 % of wall area above flashing damp at the end of 1 day. No leaks through wall at the end of 1 day.
- Class F - Water visible on back of wall above flashing in more than 3 h and less than 24 h. Rate of leakage less than 1 L/h at the end of 1 day.
- Class P - Water visible on back of wall above flashing in 3 h or less and rate of leakage less than 5 L/h at the end of 1 day. Water visible on back of wall above flashing in more than 3 h and less than 24 h and rate of leakage more than 1 L/h and less than 5 L/h at the end of 1 day.
- Class L - Rate of leakage through wall equal to or greater than 5 L/h at the end of 1 day.

The ASTM E514-74 ratings were often unofficially referred to as "Excellent," "Good," "Fair," "Poor," and "Lousy" within the industry.

Monk on ASTM E514 for Field Conditions

Monk [5] proposed a rating system that resembled ASTM E514-74, which stated that, for 10 psf (500 Pa) test pressure, water flow rates through masonry should be rated as follows:

Expected: ≤ 1 L/h

Significant: between 1 L/h and 5 L/h

Excessive: ≥ 5 L/h

ASTM E514-74 placed laboratory fabricated wall specimens with these same leakage rates in categories of "F," "P," and "L."

Hoigard et al. on ASTM E514 "Tests in Field Evaluations of Brick Masonry"

Hoigard, et al. [6] proposed the following rating system for field constructed brick masonry based on ASTM E514 tests modified for field use:

- Walls with a permeance rate below 1.9 L/h can be achieved when industry recommendations for workmanship are strictly followed and compatible materials are used.
- Walls with a permeance rate between 1.9 L/h and 3.79 L/h can be achieved in standard production masonry when industry recognized workmanship recommendations are generally followed and compatible materials are used.
- Walls with a permeance rate between 3.79 L/h and 7.57 L/h should be considered suspect. Ordinary production brick masonry construction in reasonable compliance with industry recommendations can be expected to provide better performance than this. Walls with rates of this magnitude, in situations where permeance is critical to the serviceability of the wall, should be investigated in more detail.
- Walls with a permeance rate greater than 7.57 L/h should be considered poor. Ordinary brick masonry construction can be expected to provide much better performance than this. Walls with rates of this magnitude usually result from workmanship that ignores industry recommendations, wall materials that are not compatible, or both.

Discussion of Rating Systems

Monk suggested that water permeance of field constructed masonry be rated using a system with permeance rates similar to ASTM E514-74, which was developed for laboratory constructed test specimens. This system indicates that water penetration rates greater than 5 L/h are excessive.

Hoigard et al. have opined that rates in excess of 7.57 L/h should be considered "poor" and result from "workmanship which ignores industry recommendations, wall materials which are not compatible, or both."

The data presented herein contain numerous cases where the penetration rates of field constructed brick masonry far exceed those proposed in the rating systems of the E514-74 standard (Monk [5] and Hoigard et al. [6]). Published reports by Coney et al. [7] also contain results far in excess of the discussed rating systems.

All rating systems in ASTM E514 were dropped in the 1986 edition and have not been reinstated since. The ASTM E514 rating system was dropped because of attempts by consultants to use the ratings to judge the field performance of wall systems. This was contradictory to the intended use of the test method.

By review of the BMS 7 data and other published field test data, including that reported herein, it is clear that penetration rates in the rating systems previously described are not realistically achievable in the typical construction environment for brick masonry. Based on the available data, the authors have prepared the following permeance rating system for field constructed nominal 4 in. (10.16 cm) thick clay brick masonry veneer.

$$\text{Good:} \quad \leq 10 \text{ L/h}$$

$$\text{Expected:} \quad \text{between} 10 \text{and} 15 \text{ L/h}$$

$$\text{Excessive:} \quad \geq 15 \text{ L/h}$$

The authors recognize that a rating system for water penetration resistance of a masonry wall surface cannot be used to define the functional acceptability of the total wall system. Wall systems with masonry wythes that exhibit poor water penetration resistance may function acceptably provided that the flashing systems and back-up are sound and prevent water entry to the interior and provided that the exposure of the wall system is such that it is not susceptible to freeze/thaw deterioration. Likewise, walls with low water penetration rates may perform poorly due to flaws in flashing systems or other components of the water management system.

ASTM C1601 Test Method in Practice

The majority of the data collected by our firm has come from the testing of clay brick masonry veneer as shown in Table 1. Based on this historical testing, we have developed some basic conclusions for test data obtained for brick veneer systems. At present, we do not have sufficient data to develop similar conclusions for concrete masonry, stone, or stucco systems.

Historical test results from our in-house testing of brick veneer using the procedures of ASTM C1601 indicate that the average water penetration rates on problem projects are:

- 7.54 L/h for tests conducted with no applied air pressure;
- 15.85 L/h for tests conducted at an air pressure of 5 psf (250 Pa); and
- 21.98 L/h for tests conducted at an air pressure of 10 psf (500 Pa).

We generally consider a maximum water penetration rate of 10 L/h, at a test pressure of 10 psf (500 Pa), representative of good masonry construction workmanship and an acceptable water penetration rate for field constructed brick masonry.

In addition, historical test results from testing performed indicate that the water penetration rates for repaired brick masonry can be significantly reduced by repair methods including tuck-pointing of mortar joints or a combination of tuck-pointing and application of clear water repellents. Water penetration rates for brick masonry repaired by tuck-pointing of mortar joints or a combination of tuck-pointing and application of clear water repellents can be expected to be reduced to the following levels:

- <1.0 L/h for tests conducted with no applied air pressure;
- <1.0 L/h for tests conducted at an air pressure of 5 psf (250 Pa); and
- <6.0 L/h for tests conducted at an air pressure of 10 psf (500 Pa).

Figure 5 shows test results from a project where potential repair procedures were evaluated by conducting water penetration testing as part of a prototype repair program. Testing was performed before repairs were executed to provide baseline water penetration rates for areas of the masonry veneer facade. Prototype repairs were installed at the initial test areas, cured, and then retested to assess their effectiveness in reducing water penetration through the veneer. The prototype repairs included:

- Application of a clear water repellent coating
- Partial mortar joint tuck-pointing and application of a clear water repellent coating

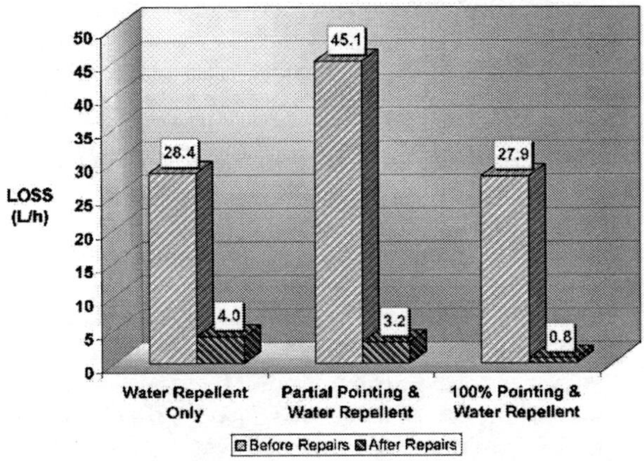

FIG. 5—*Comparison of repair method effectiveness on brick veneer using ASTM C1601.*

FIG. 6—*Water collected at flashing line during C1601 testing.*

- Complete (100 %) mortar joint tuck-pointing and application of a clear water repellent coating

Testing of potential repair methods through a prototype evaluation program using ASTM C1601 can be used to assess the increase in effectiveness of repair procedures versus increased cost of progressively more comprehensive repairs. In some cases, the increased cost of more comprehensive repair methods may not be warranted by the achieved incremental increase in water penetration resistance.

The test method can also be used to introduce water with a wall system at a rate unique to the construction to permit observation of the function of the water management system within the wall system. If water penetrating the wall system is collected and channeled back to the exterior without water entering the interior, the wall system can often be considered to be functional and may not warrant extensive repairs. Figure 6 shows a C1601 test in progress and water flowing out of the wall system at a through wall flashing line below the test chamber.

Summary

The ASTM C1601 test method is an effective tool for evaluating field performance of masonry wall systems related to water penetration resistance. The method is flexible in permitting the use of specific test conditions as well as standard test conditions that have been used in practice prior to the finalization of the standard. The test method also permits direct comparison of product and repair technique performance on in service masonry.

References

[1] ASTM C1601–06, "Standard Test Method for Field Determination of Water Penetration of Masonry Wall Surfaces", *Annual Book of ASTM Standards*, Vol. 4.05, ASTM International, West Conshohocken, PA, 2006.

[2] ASTM E514–06, "Standard Test Method for Water Penetration and Leakage Through Masonry," *Annual Book of ASTM Standards*, Vol. 4.05, ASTM International, West Conshohocken, PA, 2006.

[3] Fishburn, C. C., Watstein, D., and Parsons, D. E., Building Materials and Structures Report, Report BMS7-"Water Permeability of Masonry Walls", National Bureau of Standards, Washington, DC, October 18, 1938.

[4] ASTM E514–74, "Standard Test Method for Water Permeance of Masonry," *Annual Book of ASTM Standards*, Vol. 4.05, ASTM International, West Conshohocken, PA, 1974.

[5] Monk, C. B. Jr., "Adaptations and Additions to ASTM Test Method E514 (Water Permeance of Masonry) for Field Conditions," *Masonry: Materials, Properties, and Performance*, ASTM STP 778, ASTM International, West Conshohocken, PA, pp. 237–244.

[6] Hoigard, K. R., Kudder, R. J., and Lies, K. M., "Including ASTM E514 Tests in Field Evaluations of Brick Masonry," *Masonry: Design and Construction, Problems and Repair*, ASTM STP 1180, ASTM International, West Conshohocken, PA, 1993, pp. 314–325.

[7] Coney, W. B. and Stockbridge, J. G., "The Effectiveness of Waterproofing Coatings, Surface Grouting, and Tuckpointing on a Specific Project," *Masonry: Materials, Design, Construction, and Maintenance*, ASTM STP 992, ASTM International, West Conshohocken, PA, 1988, pp. 220–224.

Journal of ASTM International, Vol. 4, No. 10
Paper ID JAI100875
Available online at www.astm.org

David A. VanOcker, P.E.,[1] *Erin M. Johnson, EIT,*[1] *and Tracy D. Marcotte, Ph.D, P.E.*[1]

The Identification of Corrosion-Related Damage from Cramp Anchors in a Limestone-Clad Building Façade Using NDE Techniques

ABSTRACT: A recent restoration project on a 100-year-old, limestone-clad building provided an opportunity to evaluate the efficacy of certain non-destructive evaluation (NDE) techniques in identifying corrosion-related distress of embedded cramp anchors. After decades of service, progressive corrosion of the anchors had caused limestone delaminations and spalls, and presented a life-safety risk when fragments of limestone became dislodged and fell from the façade. Adding to this risk, it became clear early in the investigation that for certain areas of corrosion damage, stone cracking was not apparent on the surface or was otherwise difficult to detect. In an attempt to identify these pre-emergent cracks and delaminations, a range of NDE methods was considered to supplement the traditional techniques of visual identification and sounding. Of the NDE techniques implemented (i.e., ultrasonic pulse velocity surveys, infrared thermographic scans), infrared thermographic scans proved to be the more effective diagnostic tool to locate pre-emergent cracking and delamination sites, and showed promise as an early detection technique for similar buildings.

KEYWORDS: cramp, anchor, limestone, façade, investigation, infrared, NDE, ultrasonic, corrosion, thermography

Introduction

Transitional masonry clad construction relies upon mechanical ties to anchor a decorative veneer (e.g., dressed stone, cast stone) to a masonry backup, commonly comprised of brick or concrete masonry units, cementitious materials, or coarser, loose-laid stone. In the past, these ties were commonly manufactured from ferrous materials and historically, iron and steel ties may have been protected from corrosion by filling the remaining mortise volume with molten lead or a cementitious cover, or the ferrous material was coated in oil or galvanized [1,2]. In the northeast United States, however, it is also common to find ferrous cramps embedded in lime-based mortars with no evidence of any protective steel coating, especially if the structure was constructed between 1890 and 1920.

After decades of exposure, a lime-based mortar cures and hardens through carbonation and the environment in which the cramp is set is no longer protective. Coupled with voids in the embedment developing differential corrosion conditions (galvanic cells), the resulting build-up of corrosion products can generate sufficient internal stresses to fracture a veneer material. The fracture of the material below the surface can lead to a delamination which might stay attached to the building for some time or could fall off (spall) without warning. The process required to identify and address these deficiencies must be repeated periodically for the life of the building as the building ages and new deficiencies emerge. Guidance for repairs is available [3], but there remains the possibility of overlooking a single deficiency, leaving considerable risk.

The use of non-destructive evaluation (NDE) techniques as a diagnostic tool for historic structures in the United States is emerging, and overcomes the limitations of invasive probing of hidden construction [4]. However, the full potential of these techniques has not been realized because procedures have not been standardized for these applications and specialized personnel are needed for the interpretation of findings. Moreover, no technique applied in isolation appears to provide all needed information, and a multi-technique, multi-faceted approach is often merited. The intent of this paper is to examine the contribution of NDE techniques in diagnosing corrosion-related damage of cramp anchors for a particular case study,

Manuscript received October 21, 2006; accepted for publication September 24, 2007; published online October 2007.
[1] Principal, Staff Engineer, and Associate respectively, CVM Facilities Renewal, 85 Old Eagle School Road, Wayne, PA.

and their potential opportunities for helping building owners manage their properties in a safe and as financially efficient manner as possible.

Case Study Background and Theoretical Considerations

Overview

A university in the northeastern United States was planning a large interior renovation program for one of its Colleges, a complex of residential and academic buildings. For one of the more significant tower buildings serving as an entrance to the central quadrangle, a façade restoration program had been conducted approximately 10 years earlier, and the planned renovation program thus was not initially focused on any exterior restoration needs beyond window replacement. However, as the project planning progressed, the long-term durability of the residence hall façade was questioned.

The residence hall was built in 1905 with limestone-clad load-bearing masonry walls. The north and south wings are five stories high with gabled clay tile roofs, as shown in Fig. 1. The center section of the building is seven stories high with a flat roof bounded by four 20-ft-tall turret structures. Beyond the architectural expression of these elements, the turret walls each housed as many as 14 chimney flues, as the building was originally heated by fireplaces. With regard to the walls, the limestone cladding was laid in a random ashlar pattern, typically ranging in thickness from 4 in. to 8 in. At the upper levels, walls were as much as 16 in. to 24 in. thick at the parapets. Given the age of the structure it was considered likely that ferrous anchors were tying the construction together, and the probability of some degree of anchor corrosion was high. With this understanding, the owner of the building requested a new study to confirm the condition of the cramp anchors and the limestone cladding.

FIG. 1—*Overview of limestone-clad east façade and central turrets.*

Preliminary Façade Observations

Limestone distress was clearly visible from the ground level, with a number of delaminations and spalled areas present in the vicinity where embedded cramp anchors were likely to be present. The approximate locations are schematically illustrated in Fig. 2. Spalled limestone fragments were observed lying on the ground. To mitigate the potential for additional spalls, emergency swing stage inspections and removal of delaminated limestone fragments were conducted over a two-week period. During this period, it was established that approximately 200 limestone panels, approximately 2 % of the number of façade panels, contained delaminations or spalled areas.

Through exploratory probing, additional observations of the façade anchorage were made. The outer edges of the cramp anchors were typically observed to be bent downward into kerfs that had been hand-cut into the tops of limestone panels. The location of the legs was found to vary from as little as 1 in. to as much as 3 in. from the exterior face. Figure 3 illustrates the internal configuration. Similar to multi-wythe

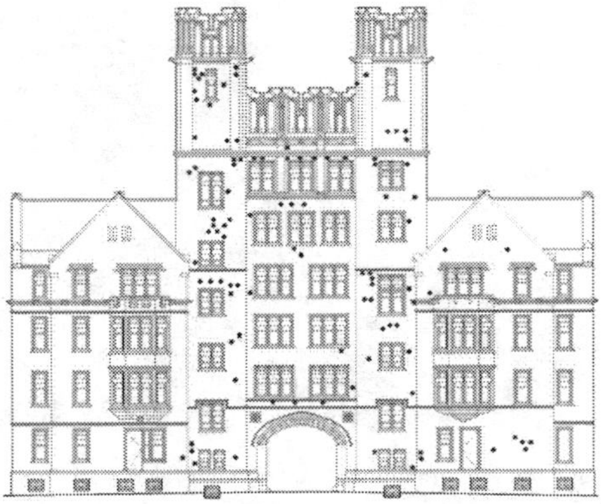

FIG. 2—*Delamination and spall locations on the east elevation.*

masonry construction, the bed joints and collar joints were filled with mortar to varying degrees, and corrosion was found to occur in varying zones along the length of the cramp anchors. Corrosion occurred in increased severity at locations where the mortar did not fully encase the anchor. Although the mortar was fully carbonated and did not provide corrosion protection to the steel (pH$<$11), a more uniform corrosion attack over the entire surface of the anchor had been anticipated. However, the presence of intermittent voids in the mortar along the anchor surface created differential corrosion environments, resulting in localized corrosion attack to the cramp anchor at the voids, where the moisture and oxygen levels were high.

The typical path for moisture penetration through the cladding is along the bond line between the mortar and stone. An unusual condition occurred at the turret levels, where rainwater flowed freely into the chimney flues. Here moisture penetrated into the back-up construction through failed joints in the terra cotta and brick flue linings, leading to corrosion of the anchors far inside the mass masonry. Complete corrosion of the cramp anchors (disintegration) was discovered in these regions with little or no external signs, such as spalling. The few signs of outward distress that did appear were related more to freeze-thaw damage and resultant differential movement in parapet and turret panels, rather than cramp anchor corrosion.

No apparent pattern was found in cramp anchor corrosion with respect to façade height, orientation, or even shielding (beneath projecting architectural elements) that could be used to correlate occurrence of corrosion to physical locations on the façade. Thus, it was not possible to discern areas where cramp anchors were more susceptible to corrosion. The only reliable way to find delaminations before they were visible from the ground was through up close surveying.

After a full presentation of these findings, the building owner was astonished, believing that their restoration program from 10 years earlier was completely successful in mitigating all potential risk. Unfortunately, cramp anchor corrosion had continued; even the patches from the recent restoration program were failing. These preliminary findings had major implications on the renovation program: both short term and long term. Short term, there were obvious safety, budget, and disruption issues; long term, there was justifiable concern over how to mitigate the recurrence of the delaminations and spalls.

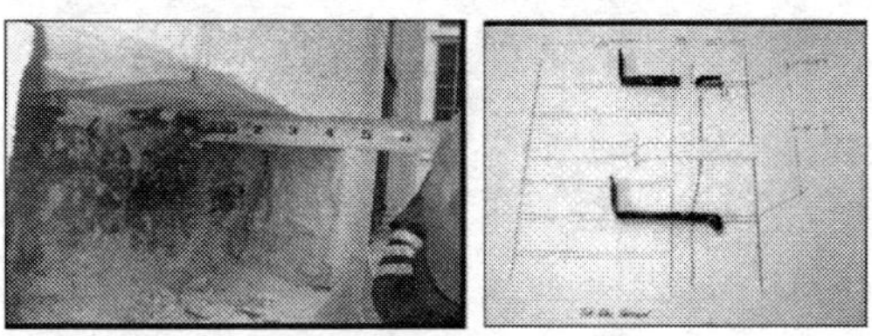

FIG. 3—*Cramp anchor location and localized corrosion of anchors at collar joints.*

FIG. 4—*Example of pervasive limestone cramp anchor spalling in major commercial buildings. Overview of the structure, left, and a closer view of a particular area, right.*

Records from the earlier restoration program were sought in an attempt to establish a trend and extrapolate future rates of limestone spalling. Unfortunately, previous inspection and restoration program records provided insufficient information to support the predictions of future rates of limestone spalling. The conservative assessment of the limited data, one that minimized the risk of further spalls, was that the spalling phenomenon was likely to increase in frequency with time, due to increased moisture infiltration and the advancing carbonation front. It was concluded that another major façade restoration should be implemented immediately.

Preliminary Owner Recommendations

A range of recommendations for restoration and life safety programs was considered. One commonly utilized technique is to access the building via swing stage once a year, to identify and remove the delaminations, deferring formal spall repairs to 5-year intervals. Figures 4 and 5 show similar buildings in a densely populated city where such a reactive strategy has been employed since the 1980s. This approach does not involve any efforts to slow or arrest the cramp anchor corrosion, the route cause of the limestone delaminations and spalls. Such a strategy requires annual inspections for the life of the building.

The owner also was presented with an option to implement a cathodic protection system based upon galvanic anodes. The system would have electrically connected galvanic anodes to the embedded steel cramps to reduce their corrosion rate and the subsequent spalling of the limestone. This technology has been successfully used to mitigate the corrosion of steel in concrete and is generally accepted, but its use in this application was considered new and experimental. Another concern was that the galvanic anodes would be consumed with time and would require replacement. Further testing would have been required to estimate the anode consumption rates and guarantees of no further limestone spalls in the owner's ideal

FIG. 5—*Overview of a building with similar limestone façade construction and vintage to the case study building, left. A closer review of one area with the multitude of Dutchman repairs that represent the magnitude of cramp anchor induced distress, right.*

service life of 40-50 years were not possible. There has been more work to date in the area of cathodic protection systems that rely on impressed current: these systems are designed to electrically connect all components requiring protection using a network of wires routed through the mortar joints, install a non-consumed anode array and reference electrodes, and apply a current that counterbalances the corrosion current. A successful application of this technology to a limestone façade has been summarized by Jones and Ballard [5]. The main concern with this strategy for the residence hall of this paper was that any future maintenance repointing efforts could damage the wires in the mortar joints and mandate another complete installation.

The owner chose to mitigate future delaminations and spalling by slowing the rate of cramp anchor corrosion through a restoration program that included 100 % repointing and applying a silane-siloxane based sealer to render the joints resistant to further ingress of moisture. Limestone panels containing cramp anchor distress were replaced completely. The chosen program option further included an ongoing maintenance budget for future inspection and spall removal at 3- to 5-year intervals, followed by full repair every 15 to 20 years. The investigative team was charged with developing an evaluation program, to be conducted during the restoration period, that would provide a high level of assurance and safety during the interval between major restoration cycles.

Project Goals and Evaluation Strategy

Goals for the project initially were to identify delaminated and spalled locations for immediate repair, and to establish cost-effective and efficient measures to slow the rate of corrosion, given the likelihood that anchors would continue to corrode at an uncertain rate. As further pilot exploration and evaluation phase work progressed, another goal became apparent: the need to identify pre-emergent subsurface conditions that would eventually lead to cracking and spalling, but that had not developed to a point that could be readily identified through normal survey techniques. This added goal was based on the desire to obtain a higher level of assurance that new delaminations and spalls would not occur for as many years as possible following the completion of this new restoration program. The goal is to provide the owner with a higher degree of certainty of the safety of the façade over a longer period of time. This final goal inspired the consideration of NDE techniques to enhance our traditional methods of detecting pre-emergent cracks.

In combination with traditional visual observation and sounding techniques, several NDE methods were considered during the project with regard to their feasibility to serve as viable methods for owners, design professionals, and contractors to employ in the future. These traditional and NDE methods are described in subsequent sections. Ultimately, time constraints of the construction schedule permitted only two NDE techniques to be field tested: ultrasonic pulse velocity (UPV) and infrared thermographic (IR) imaging. The restoration program involved complete dismantling and reconstruction of one of the turrets, and replacing the brick backup (and flues) with reinforced concrete masonry construction. This provided a perfect opportunity to conduct a series of non-destructive tests to detect subsurface anomalies (cracks) and then to inspect conditions as the masonry was deconstructed. The contractor exercised great care in removing the limestone; per the owner's requirements, panels without delaminations or spalls had to be salvaged and reused 100 %. Removals typically proceeded in a slow methodical fashion that enabled careful inspection of the cramp anchors in each piece as the walls were taken apart. Results of the UPV and IR testing are presented in the Results and Discussion section.

Traditional Delamination and Spall Identification Techniques

Visual inspection and sounding are the most common techniques for detecting delaminations in masonry and concrete materials. It is common that visual inspections alone are relied upon to detect such conditions. In many cases visual inspections are conducted with binoculars from the ground, from low roofs of the building and from windows, balconies, and roofs of neighboring buildings. However, visual inspections from a distance can be unreliable and may not provide confidence that safety hazards do not exist. Visual inspections from afar are inadequate to detect faint or emerging cracks, thus in many cases the full extent of delaminations cannot be reliably detected from the ground, even with binoculars. Crack orientation (i.e., inclined upward rather than downward) may be an impediment to detection, as correlation to the observer's viewing angle is directly related to whether the crack is noticeable. Examples of varying crack orientation are shown in Fig. 6. Horizontal projections and shadow lines can occlude the presence of

FIG. 6—*Examples of delaminations that are difficult to detect without close inspection.*

cracks. Cracks and delaminations on the backside of the limestone panel are impossible to identify with visual inspection. In our case study approximately 35 % of the fractured or delaminated limestone panels were discovered via visual inspection with binoculars from afar.

Sounding is very effective in determining whether a crack in a limestone panel is leading to a delamination, but is labor intensive and up-close access to the area is required. During the pilot phase of the case study, sounding was performed with a mason's hammer and rubber mallets, exercising care to avoid marring the surfaces. During this process, though, a considerable additional concern was raised when some of the panels that exhibited no visible evidence of cracks or delaminations suddenly fractured when the mortar joints were sawcut. After this incident, sounding and up-close visual observations were conducted after the limestone mortar joints had been sawcut. In our case study, the other 65 % of the cracked and delaminated limestone panels, located using traditional techniques, were identified through up-close visual examination and sounding. While sounding and up-close visual observations appeared to be effective, we still felt that there had to be more efficient, less invasive methods to detect fractures that were likely originating deeper inside the stone.

Ultrasonic Pulse Velocity Testing

Testing in accordance with ASTM C 597-02, "Standard Test Method for Pulse Velocity Through Concrete, Ultrasonic Pulse Velocity (UPV)," is typically performed as a direct test through a concrete surface (i.e., a transmitter is placed on one side and a receiver on the other side); this is referred to as Direct Mode UPV [6]. The transmitter sends out a stress wave that is received on the other side, with the speed of propagation being measured. The stress wave naturally takes the most direct, fastest, path through sound concrete. When the stress wave encounters a defect in the concrete, the wave takes an alternate path around the defect, resulting in a measurably slower speed. For concrete applications, the stress wave frequency used is typically 54 kHz; this roughly corresponds to a wavelength of 3 in. and represents the minimum defect size that can reasonably be detected. Physical properties of the concrete are established through destructive tests on core samples, and correlated with the path time of the stress wave to estimate the compressive strength.

For this façade case study, as only one side of the limestone panels was accessible, a modified test method known as Indirect Mode UPV was used. This method has been used to study concrete pavements and structures. Because the limestone has a different density and elastic constant than the grout filling the collar joint, every limestone panel is effectively isolated from the others. Therefore, the test can be performed from two different locations on the same surface. For the Indirect method, the transmitter and receiver were placed on the same surface separated by a known distance; one measurement was taken at the bottom of the panel, presumably where the limestone is defect free, and the other was taken at the top where any corrosion of a cramp anchor may have initiated a crack. This technique is illustrated schematically in Fig. 7. The working assumption was that a difference in travel times between the test at the top and the test at the bottom may indicate the presence of a crack in the panel. These measurements were compared to the actual conditions observed during the selective demolition and dismantling of one of the turrets, to establish the accuracy of this technique. Overall, 37 panels were studied within five wall areas. Along with some calibration tests conducted on a representative limestone panel, these UPV results are presented at the beginning of the Results section. The presented results are compared using their compressive strength values calculated from the calibration data. This extra step was necessary to harmonize the varying path lengths (i.e., typically 12 in., but ranging from 8 to 14 in.).

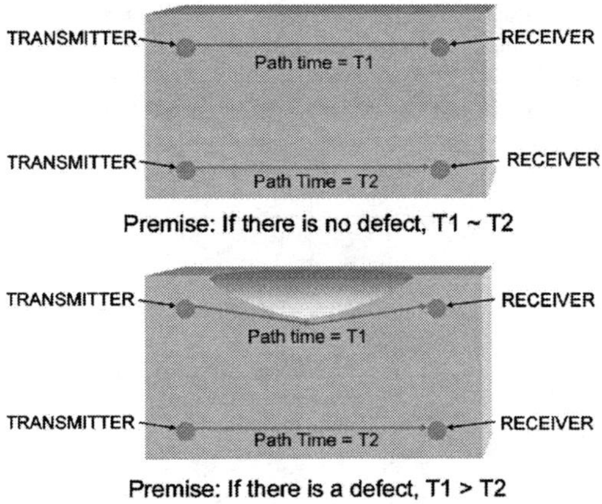

FIG. 7—*Schematic illustration of the Indirect Mode UPV used to investigate any damage from corroding cramp anchors.*

Infrared Thermographic Imaging

By adapting ASTM C 1153, "Standard Practice for Location of Wet Insulation in Roofing Systems Using Infrared Imaging," the practice of using infrared thermographic (IR) images has been extended to use as a diagnostic tool for locating façade deficiencies. The technique is summarized by Kaplan et al. and in ASTM C 1153 [4,7]. In general, discontinuities in facades, such as delaminations, affect heat flow through the material and result in localized differential surface temperatures. In addition, precipitation or significant applied moisture is understood to have a significant impact on the images and it is commonly used as an indicator of leak points in building envelopes. For this case study, it was speculated that IR images could detect faint cracking if the testing were conducted shortly after rainfall, under the premise that moisture that is trapped in cracks evaporates more slowly than from the surrounding intact limestone surfaces. It was also thought that the IR images would be more effective if they were taken when changes in building surface temperature would be most prevalent and the differences in heat flow at a discontinuity would be the most noticeable. IR images were collected for several elevations when the surface temperature was rising or falling (e.g., early morning, late evening), and shortly after a light rainfall; these results are presented in the Results Section, "Infrared Thermographic Scans."

Impact-Echo Techniques

Impact-Echo techniques have been established as viable methods to determine the thickness of and detect flaws (cracking, etc.) in concrete [8]. An advantage of the impact-echo method for this project was that access to only one side of the limestone would have been needed. Others have reported success with the technique working with homogenous stone materials (i.e., granite and sandstone columns, brick walls, veneer walls) [4]. However, the one drawback that led to the decision not to implement this method for our case study was that the limestone finish did not provide a sufficiently flat surface. The crandalled surface finish had a roughened texture with the difference between locally high and low spots on the stone approximating 1/8 in. Therefore, consistent, good contact between either the mechanical impactor or the receiver with the limestone surface could not be assumed.

Results and Discussion

Ultrasonic Pulse Velocity Measurements

Preliminary Calibration Standards—Prior to implementing any field testing, a representative limestone panel from the building was studied to provide calibration data for the subsequent indirect UPV field measurements. This stone was also studied petrographically and its macrostructure appeared to be consistent with Indiana Limestone. Table 1 summarizes compressive strength and bulk density measurements performed on two core samples (both 3 in. diameter by 4 in. long) taken from the unit. The results agree with published values for Indiana limestone [9].

TABLE 1—*Summary of control limestone core tests (Courtesy of Valley Forge Labs).*

Core sample	Compressive strength (psi)	Bulk density (lb/ft^3)
A	6640	146.2
B	7090	147.2
Mean	6865	146.7

The conversion of UPV data to provide reliable quantitative compressive strength results is limited to Direct Mode UPV data; therefore, the control panel was measured by UPV in both the Direct and Indirect Modes. The input of the density and compressive strength test results from the core samples is necessary to calculate the UPV apparent compressive strength. With this physical property data, two coefficients were calculated that permitted direct mode UPV path times and lengths data to be used to correlate the measured path times and lengths to the mean compressive strengths. These data were then estimated to provide reliable quantitative compressive strength values that are summarized in Table 2. These calculations can be compared with UPV data collected using the Indirect Mode measurements planned for the field. As such, an Indirect Mode apparent strength value of 2562 psi statistically correlates to a Direct Mode calculation of 6892 psi. This correlation was considered reasonable because of the study conducted by Yaman et al. [10] on concrete. This study validated the statistical relationship between Indirect and Direct UPV in concrete, a material far more heterogeneous than limestone. Therefore, if the correlation between Indirect and Direct UPV for concrete is statistically valid, then the correlation for limestone can be assumed.

TABLE 2—*Summary of calibration of Direct to Indirect Mode UPV measurements (Courtesy of Valley Forge Labs). Assumption: Poisson's Ratio = 0.14.*

UPV Mode	Path time (μm)	Path length (in.)	Calculated apparent strength (psi)	Correlation Values used for field investigation (psi)
Direct	93.8	14.75	7258	6892 (Mean)
	38.4	5.88	6526	
Indirect	66.0	8.00	2562	2562

Field Results—For the 37 limestone panels tested, two Indirect Mode UPV measurements were taken per panel, one at the top of the panel near the cramp and the other at the bottom; results were paired to compare any differences. These paired results are presented in Fig. 8. A reference line is marked representing the assumption if no defect is present; the top of the panel result should roughly equal the bottom panel result. A general review of the data indicates that even if the trend line were broadened to provide a confidence interval, there is little, if any, precision in the results, and it is difficult to establish any trends in the data. However, there is an indication that there is a typical limestone quality, referred to as "medium-density limestone" in ASTM C 568, and the strength approximations are consistent with the published data for Indiana limestone [9]. In addition, certain limestone panels appeared to be significantly stronger and could be considered "high-density limestone" by ASTM C 568 [11]. This finding was supported by follow-up destructive laboratory testing. The Indirect Mode UPV was, therefore, successful in distinguishing between varying grades of limestone.

With regard to locating internal defects, it appears that the technique was not as effective as when it is used to investigate concrete. In the area below the reference line, results in this region could indicate that the transit time relatively increased on the top of the panel, and a crack or defect might be present. Overall,

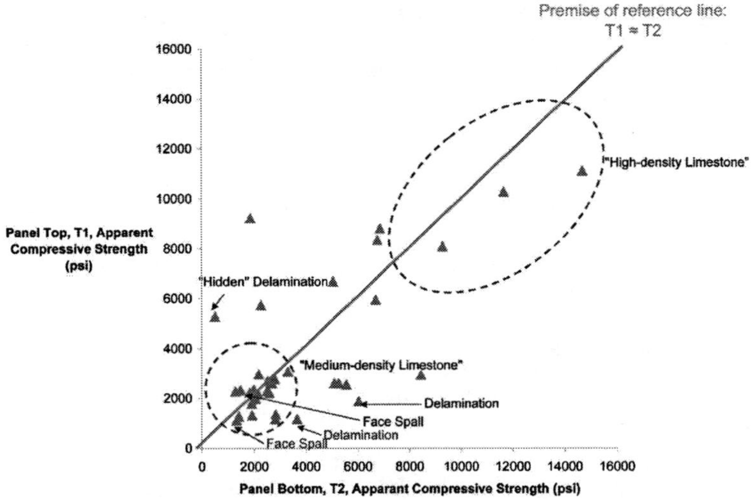

FIG. 8—*Comparison of the paired Indirect Mode UPV results.*

at least seven results were measured to have a significant difference that could indicate a defect. However, only two of the seven panels were discovered to actually have a defect when they were removed from the façade. These panels are labeled "delamination" in Fig. 8. Complicating this interpretation, the two panels with delaminations do not appear to have a statistically significant change in transit time (i.e., the calculated compressive strength for the top of the panel in the region of the anchor is approximately the same as the bottom measurement). In addition, another panel, labeled "hidden delamination," contained a defect that was not revealed until the mortar joint was sawcut, and this result was not even located in the region where defects were anticipated to be observed in Fig. 8 (i.e., below the reference line). It can be speculated then that prior to the sawcutting interaction with the panel, there may not have been a crack, but corrosion was generating internal stresses. With the energy of the sawcutting action, a critical threshold may have been reached, resulting in the spall. Overall, though, it is clear that for this and the other defective panels within the 37 panels studied in total in this investigation, the sample pool was too small to draw firm conclusions. In addition, the possibility of "false positives" suggests that the technique might not be more efficient than traditional techniques.

The almost equal number of paired samples below the reference line represents scatter in the data that must be understood before confidence in the results can be achieved. The scatter in the data could be attributed to the following:

(a) Defects unrelated to the cramp anchor may be present at the bottom of the panel.
(b) Although well suited for concrete studies, the frequency of the applied stress wave may be too low for limestone investigations, as the corresponding wavelength might be too large (~2.3–2.9 in.). Other transducers with frequencies as high as 150 kHz are readily available and are often used in ceramic studies. For the limestone in this investigation, the use of a 150-kHz transducer would lower the wavelength of the stress wave to under 1 in., effectively increasing the resolution of the technique.
(c) The bedding planes of the limestone add an additional variable that is not encountered in concrete studies. The impact of bedding planes on the results might be better understood if more than two measurements were taken per panel, creating a profile of the surface. However, this would greatly undermine the speed of the technique and would become a disadvantage.
(d) The irregularity of kerf cut into which the cramp anchor was placed could impact the transit time of the wave.
(e) It is not possible to know whether the thickness of the stone varies from top to bottom without exposing the head joints. While typical small variations of approximately 1/2 in. did not appear to affect the data, there were some stones found to have greater than a 2 in. variation in thickness

from top to bottom. The larger thickness variations lead to a greater travel time, without one noticing that the path length is larger. In consideration of all these experimental uncertainties, it can be reasonably concluded that much more research is required before this method can be used with confidence in field applications.

Infrared Thermographic Scans

In accordance with the strategy described in Project Goals section, "Infrared Thermographic Imaging," testing was performed at night and in the morning at different locations on the façade of the case study building. During the evening observations, a cold front with associated rain showers passed through the area prior to testing, providing ideal conditions for IR imaging. Air temperatures had dropped into the low 60s (°F), while the building surface temperature was still in the 70s (°F). As shown in Fig. 9, areas with clearly spalled and delaminated limestone panels were distinguishable. This area had been identified during our initial surveys described in Case Study section "Preliminary Foçade observation," and served as a starting point for optimizing the technique. Immediately, it became clear that a strong advantage of the technique is that information can be reviewed in real time, but also recorded and stored for later review.

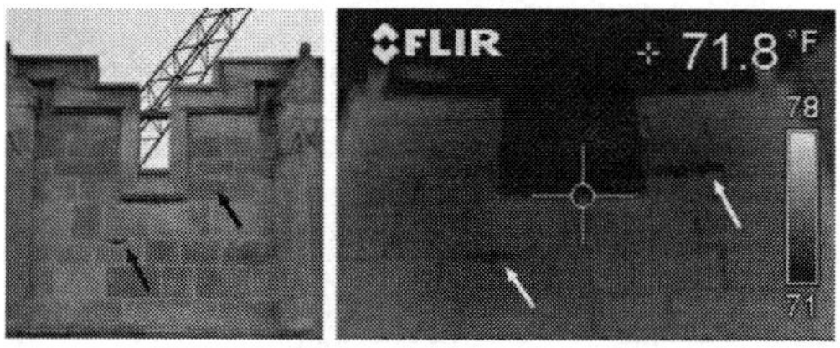

FIG. 9—*Clearly visible spalled areas, left, were readily identified with IR scans, right.*

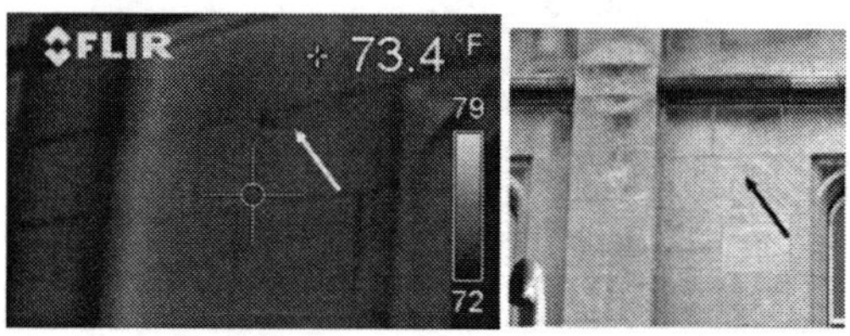

FIG. 10—*IR scan, left, revealed a hidden delamination from 10–20 ft away that could not be observed with the unaided eye, right, at a similar distance.*

In other locations, delaminations could be detected 10–20 ft away using IR, that were much more difficult to see with the unaided eye from the same distances. There was also a clear thermal signature of the panels that had previously been replaced, likely due to differences between new and old materials having a combination of different emissivity (infrared radiation), and differing density, which will lose heat at a different rate. This was validated by comparing the locations on the IR images to the record document depicting the replaced limestone panels from the ongoing construction phases. As another example, Fig. 10 (left) shows a small hidden defect that was picked up by the IR scanner from approximately 10 ft away. This defect did not show up on the corresponding photograph taken from the same distance, presented as Fig. 10 (right). In fact, this crack could not be detected with the unaided eye until the surveyor was a few inches away from the location. Thus, the IR scanner was able to show delaminations and other anomalies in the limestone surface that would probably have been otherwise missed.

One unexpected observation that was noted with the IR scanning was the ability to observe failed joints in the areas which had yet to be cut out and repointed, as shown in Fig. 11 (left). The joints were not

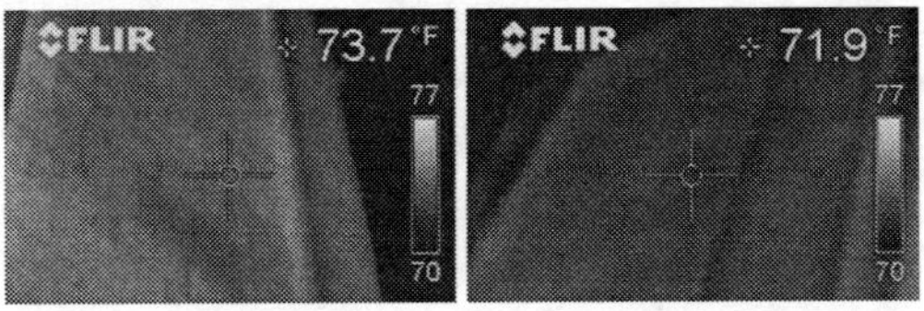

FIG. 11—*Failed joints in mortar, left, and a repointed wall, right.*

FIG. 12—*Failed mortar joint.*

visible in thermal images of the areas that had already been cut out and repointed, as shown in Fig. 11 (right). The failed joints are areas where there is no longer a bond between the mortar and the limestone panel as shown in a close up view in Fig. 12.

As mentioned previously, IR is very sensitive to weather conditions, and this was exemplified by the results of the morning investigation during which the air temperature did not warm up, and the surface and air temperatures were almost identical. We were unable to obtain any useful information during those thermal scans, similar to the experience reported by Kaplan et al. [4]. Only the areas in direct sunlight showed differences in surface temperatures. These areas, however, showed varying surface temperatures, not due to anomalies, but due to shadows from neighboring trees, building projections, and scaffolding. Therefore, it can be concluded that IR scans can assist in providing fast, insightful observations of corrosion-induced damage by cramp anchors in limestone façades, but that the timing of the observations is critically important.

Conclusions and Recommendations

1. Traditional techniques of close-up visual inspections and sounding are extremely effective in locating most delaminations and spalls in stone cladding; however, this process is time consuming and latent or emerging delaminations can be missed.
2. UPV and IR NDE methods can enhance the information collected from visual and sounding surveys, as hidden deficiencies were identified.
3. Infrared thermographic scanning was extremely successful in locating defects in limestone panels from a distance of 10–20 ft that were difficult to observe with the unaided eye from only a few inches away. Although weather and temperature conditions must be considered, IR scanning techniques appear to have a great deal of promise for future use in building façade evaluations with cramp anchor concerns. Use of this technology could provide valuable information into the condition of a façade without the need to scaffold the building. Tests could be performed from the roof of a neighboring building or from a high reach some distance away. Further evaluation of this technology should be pursued to determine its applicability in different façade applications.

4. Indirect Mode UPV testing was able to detect differences in limestone grades.
5. Indirect Mode UPV provided inconclusive results regarding cramp anchor corrosion-related defects for the limestone panels studied, although it did identify one limestone panel with a hidden deficiency. The lack of precision in the data would merit considerable destructive investigation to validate the results. Were it not for the opportunity of reviewing the panels while the turret was being dismantled, a number of false positive potential defects would have been identified, representing a considerable waste of resources and time. Thus, more data should be collected and assessed before the effectivemess of indirect mode UPV can be fully understood.
6. Indirect Mode UPV also highlighted the need for further research to continue to adapt this concrete technology technique to limestone applications: (a) the potential use of higher frequency transducers (possibly as high as 150 kHz), and (b) determining the influence of bedding planes and other panel geometries affecting the transit time of the stress wave.
7. Although it is too early for any confirmation, the results of this study suggest that when investigations enhanced with traditional and NDE observation techniques are coupled with a restoration program tailored to address the observed conditions, the development of a lower risk, more cost-effective management program may be achievable.

Acknowledgments

The authors would like to thank Jonathan Hunt and Michael Boyle of Valley Forge Laboratories of Devon, PA, for their assistance in developing the testing program, and collecting and interpreting the ultrasonic pulse velocity measurements. In addition, we are also grateful for the assistance of Paul Spiegel of Construction & Materials, LLC of West Chester, PA, who collected the Infrared Thermographic imaging scans.

References

[1] Elliot, C. D., "Iron and Steel," in *The Development of Materials and Systems for Buildings, Technics and Architecture*, Massachusetts Institute of Technology, Cambridge, MA, Second printing, 1993.

[2] Warland, E. G., *Modern Practical Masonry*, Donhead Publishing Ltd, Dorset, UK, 2006 Ed., Reprint 1929.

[3] Pieper, R., "Preservation Brief 42: The Maintenance, Repair and Replacement of Historic Cast Stone," National Park Service, Department of the Interior.

[4] Kaplan, M. E., Ennis, M., and Meade, E. P., "Masonry Number 4: Non-destructive Evaluation Techniques for Masonry Construction," Preservation Tech Notes (PTN 40), September 1997.

[5] Jones, G., and Ballard, G., "When History Needs a Helping Hand from the Present: An Overview of the Preservation of a Historic Façade in London," *Concrete Repair Bulletin*, Vol. 20(3), 2007, pp. 16–21.

[6] ASTM, Standard C 597–02, 2002, "Standard Test Method for Pulse Velocity Through Concrete," *Annual Book of ASTM Standards*, ASTM International, West Conshohocken, PA.

[7] ASTM, Standard C 1153–03, 2003, "Standard Practice for Location of Wet Insulation in Roofing Systems Using Infrared Imaging," *Annual Book of ASTM Standards*, ASTM International, West Conshohocken, PA.

[8] Carino, N. J., "The Impact-Echo Method: An Overview," *Proceedings of the 2001 Structures Congress & Exposition*, American Society of Civil Engineers, Peter C. Chang, Ed., May 21–23, Washington, D.C., 2001.

[9] Indiana Limestone Handbook, 1998, 20th Ed., Indiana Limestone Institute of America, Inc, Bedford, IN.

[10] Yaman, I. O. et al., "Ultrasonic Pulse Velocity in Concrete using Direct and Indirect Transmission," *ACI Mater. J.*, Vol. 98, No. 6, 2007, pp. 450–457.

[11] ASTM, Standard C 568–03, 2003, "Standard Specification for Limestone Dimension Stone," *Annual Book of ASTM Standards*, ASTM International, West Conshohocken, PA.

Journal of ASTM International, Vol. 4, No. 10
Paper ID JAI100812
Available online at www.astm.org

Mark K. Schmidt[1] *and W. Robert Hannen*[2]

Snap, Crackle, Pop: Remediation of a Noisy Curtain Wall

ABSTRACT: This case study presents an investigation of disturbing noises emanating from the curtain wall system of a multi-story office building located in the southeastern United States. Shortly after the initial construction of the building, building occupants complained of loud popping noises emanating from the exterior building facade. These noises were of a frequency and magnitude to negatively influence the work environment in adjacent office space. The cause of the noises was investigated by several groups with conflicting results. At the request of the building's general contractor, an investigation was undertaken to determine the cause of the disturbing noises and to develop remedial recommendations. This investigation included a detailed review of the curtain wall shop drawings and on-site observations to identify questionable details possibly responsible for the noises. The most probable cause of the noises was identified as diurnal solar gain that produced unaccommodated movements in the curtain wall system. The orientation and curvature of the wall combined to allow a thermal expansion that progressed along the wall. The details of the curtain wall construction restrained the movement, leading to sudden release of confining stress. To confirm this hypothesis, affected areas of the curtain wall were instrumented with accelerometers and deflection gages strategically located to pinpoint the origin of the noises. To further confirm the point-of-origin identified by the accelerometers, a series of exploratory openings were made to review curtain wall details. After the source of the noise was positively identified, remedial recommendations to mitigate the disturbance were implemented by the building owner.

KEYWORDS: curtain wall, noises, thermal movements, instrumentation

History of Condition

The subject structure is a four-story 450 000 square foot office building, which was occupied in the spring of 2003. In the spring and following summer, a consistent impact-like, popping noise was reported on the south-facing radiused aluminum curtain wall. This noise was believed to emanate from the head condition at the fourth floor, where the curtain wall is connected to precast concrete spandrel panels. Some of the most disruptive noises were occurring in the fourth floor conference room at the eastern end of the radiused curtain wall.

The sounds were more noticeable and frequent on clear sunny days, indicating some correlation with radiant thermal conditions. As the eastern morning sun tracked against the western most portion of the curved curtain wall, noticeable popping sounds tracked from west to east. As the sun moved toward the west, the easternmost portion of the wall along the conference room reportedly experienced the most noise activity and intensity. Most noise episodes coincided with early morning and late afternoon periods when some portions of the curved wall were in the sun and some were in the shade. During the most intense periods, popping noises were documented approximately every 5 min and sometimes in a more rapid succession of several pops at a time. Overcast and partly cloudy days resulted in the elimination or significant reduction of objectionable sounds. Conditions also appeared to be worse during the summer months.

From August through November 2003, the general contractor, the aluminum installer, the aluminum manufacturer, the architect, structural engineer, and a building envelope consultant conducted investigations and made minor adjustments to the aluminum system. These investigations included visual inspections of both interior and exterior curtain wall conditions, temperature readings taken on the exterior and interior trim components, removal of window treatments and interior snap-on mullion covers, a mock-up thermal test, and loosening bolts connecting the curtain to upper precast concrete panels. As a result of these efforts, the noises lessened in frequency and intensity, but were still present and the definitive source

Manuscript received September 21, 2006; accepted for publication August 9, 2007; published online October 2007.
[1] Unit Manager and Senior Consultant, Wiss, Janney, Elstner Associates, Inc., 330 Pfingsten Road, Northbrook, IL 60062.
[2] Consultant, Wiss, Janney, Elstner Associates, Inc., 330 Pfingsten Road, Northbrook, IL 60062.

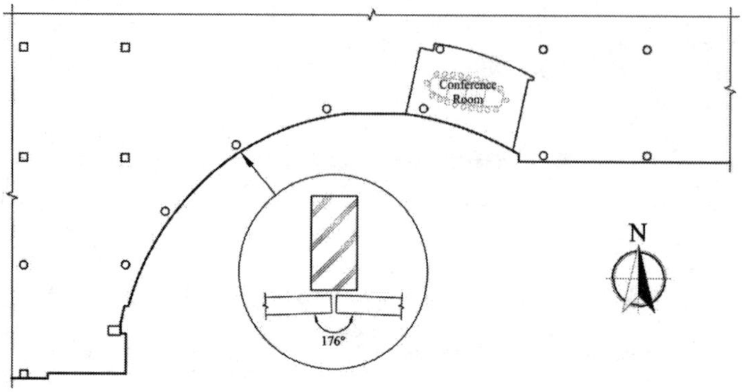

FIG. 1—*Plan sketch of curtain wall facade and schematic detail of mullion connection.*

of the remaining noises could not be identified. A mock-up test of a portion of the system was also performed by the contractors with no conclusive results.

Description of Radiused Curtain Wall

The facade of the building is a typical glass and aluminum curtain wall system adapted for use as a segmented concave wall. According to the manufacturer, previous installations of this system in concave radiused configurations had not produced objectionable noises. A plan view of the curtain wall layout for this building is shown in Fig. 1.

The curtain wall consists of alternating banded groups of vision and spandrel glass captured in clear anodized aluminum framing. There are three bands of spandrel glass and two bands of vision glass per floor. The uppermost vision glass at each floor is a short band of transom glass situated below the floor slab. A combination of exposed aluminum mullions and butt glazed glass joints exist at the vertical edges of the glass units. The precast concrete spandrel panels at the roof level are nominally 6-1/2 inches thick and covered with a foil-faced rigid insulation board. A portion of the radiused curtain wall is shown in Fig. 2.

The vertical mullions are attached at each floor slab with T-shaped aluminum anchors, which are relatively flexible in the plane of the wall. A vertical mullion stack joint occurs at the third floor level. At the head of the fourth floor vision glass, the mullions are connected to precast concrete spandrel panels at the roof level using bolted steel clip angles (see Fig. 3). Compared to the mullion anchors at the lower floors, this anchorage is significantly stiffer in the plane of the wall.

Between each horizontal rail and vertical mullion is a bolt whose head fits into a slot in the horizontal rail. The shank of the bolt penetrates the vertical mullion and is secured with a nylon locknut (see Fig. 4).

FIG. 2—*Portion of radiused curtain wall showing sunshade element (arrow).*

FIG. 3—*Upper connection of curtain wall mullions to precast concrete panels.*

By design, this connection is intended to slip to accommodate thermal movements. To create the concave radius, the angle between adjacent horizontal rails is approximately 176° as shown in Fig. 1.

Interior trim covers are all snap-on types using plastic clips and are fit tight, leaving no gaps at the intersections. Exterior covers are also snap-on type; however, small gaps have been provided for expansion and drainage. A dark painted break metal trim piece mounted to the precast concrete panels at the top of the curtain wall exists as a decorative sunshade element (see Fig. 2); it is not connected to the curtain wall.

The conference room has roll-down fabric shades for sun control, while the balance of the windows in the curtain wall have aluminum blinds mounted in the blind pocket of the head trim. The conference room also has fabric drapes for further light control during presentations.

Review of Shop Drawings and Related Standards

A brief review of the curtain wall shop drawings revealed that the most likely location for restrained thermal movement and subsequent slippage was the bolted connection between the horizontal rails and vertical mullions as shown in the schematic detail in Fig. 1 and pictorially in Fig. 4. The analysis of the drawings also revealed that the perceived popping noises occurred near location where significantly higher thermal restraint was afforded by the roof level mullion anchorage to the precast concrete panels shown in Fig. 3.

It is generally accepted that a sound level increase of as little as 3 dB above background could be detectable. Sound levels 5 dB above the background level could be considered disturbing [1].

Field Testing

The field testing was focused primarily on the large fourth floor conference room, which had particularly loud noises and comprised about one-quarter of the length of the radiused curtain wall at its east end. A basic assumption was made that the disturbing noises originated within the curtain wall as a sudden release

FIG. 4—*Connection of horizontal rails to vertical mullions.*

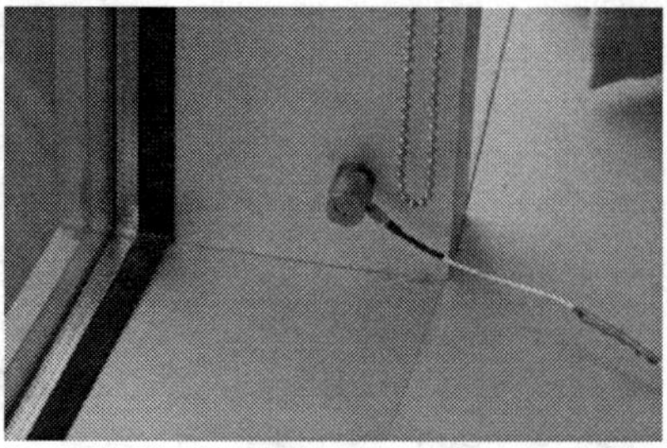

FIG. 5—*Accelerometer mounted at the base of a vertical mullion.*

of built-up stress, which produced a dynamic energy wave in the wall system radiating away from the source. The initial instrumentation consisted of eight accelerometers connected to an eight-channel dynamic signal analyzer. This system was used to triangulate the source location of the impulsive vibration. The monitoring setup also included an integrating sound level meter connected to a laptop-based dynamic data acquisition system, which stored the meter output of the continuously measured sound levels. A typical accelerometer installed on a vertical mullion is shown in Fig. 5.

The sound level meter was positioned near the middle of the conference room facing the windows. The accelerometers were attached with petrol wax to the interior snap-on covers of the vertical mullions and horizontal rails in an array designed to allow location of the vibration source through triangulation. Two different accelerometer arrays were used during this portion of the testing. The analyzer was programmed to automatically capture the sensor signals of all eight accelerometers if a trigger level was exceeded by a selected sensor.

This system successfully captured small impulsive vibrations and noises, demonstrating that the popping noises were definitely related to impulse vibrations in the framing system. However, the system required continual hands-on operation to store captured vibration signals. Since the occurrence of the problematic noises could not be reliably predicted, a second, more adaptable, system was installed to monitor continually for 2 days in an unattended manner.

The second system used a laptop-based dynamic data acquisition system to continuously monitor the output of the eight accelerometers and the sound level meter at a scanning rate of 8192 scans per second. The measured acceleration signals and sound levels were evaluated by the software every second and the peak values for each channel were stored in a log file. If any of the accelerometers received peak accelerations exceeding a set trigger level, the entire waveforms for all sensors was stored in event files for further analysis. The data acquisition system setup to record 16 channels of accelerometer data is shown in Fig. 6.

The sensors were arranged in two setup arrays, identified in Fig. 7 as Setup 1 and Setup 2. This figure shows two schematic elevation views of the window frame system in the conference room with the east end of the radiused curtain wall at the left edge of each sketch. Note that the mullions are numbered from right to left and that Mullion 2 corresponds with a column in the building. In Setup 1, each gage was

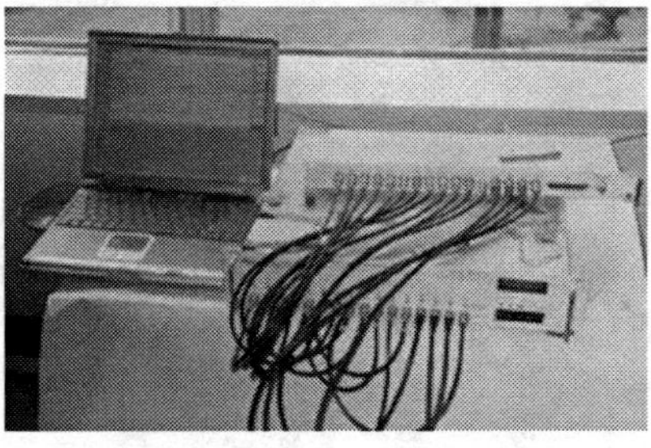

FIG. 6—*Laptop-based vibration monitoring system for 16 accelerometers.*

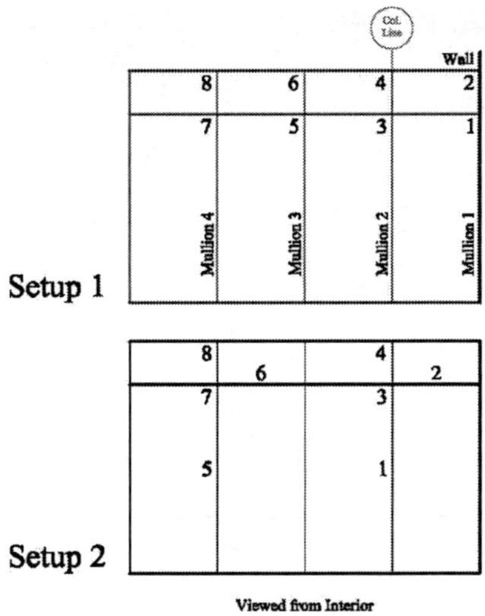

FIG. 7—*Accelerometer locations and mullion designations.*

attached to the snap-on covers of the vertical mullions measuring horizontal accelerations in the plane of the wall. After approximately 4 h of monitoring, Gages 1, 2, 5, and 6 were repositioned as shown for Setup 2.

In addition to the vibration and noise measurements, static deflection measurements were obtained at two adjacent mullions just west of the conference room. Dial gages were used to measure the relative movement between the vertical mullion and the horizontal rails on each side of the mullion as shown in Fig. 8. The deflection data was recorded manually.

Field Test Results

Figure 9 shows the peak accelerations and sound levels measured by this system in the period from 9 a.m. to midnight on one day. That day was clear but cool (less than 70°F). Many popping noises were observed and recorded by the monitoring system, but none was as large as the events that prompted the investigation. The data shown in Fig. 9 clearly illustrate the gradual buildup of impulses. As the magnitudes of the impulses increase their frequency of occurrence also increased. Then, after a quiet period, the impulses resume. Sundown triggered some relatively large magnitude events followed by a decrease in event magnitude and frequency throughout the night.

FIG. 8—*Deflection measurement apparatus.*

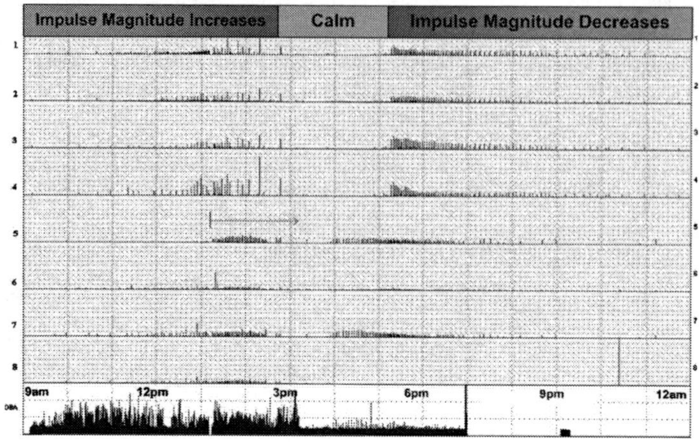

FIG. 9—*Peak accelerations from eight accelerometers and sound levels.*

The deflection data demonstrated that the horizontal rails moved toward the east relative to the mullion as much as 0.03 in. during the sunny period of the day and moved back toward the west during the afternoon and evening. The movement began at approximately the same time as the beginning of the impulse, or noise, events.

The visual examination of the curtain wall framing system led to the hypothesis that the impulsive noises were produced by releases of confining stress by overcoming the friction between the vertical mullions and the horizontal rails. The vibration log plot shown in Fig. 9 provided supporting evidence confirming that hypothesis. The question remained: Which level of horizontal rails was producing the slippages? Therefore, the analysis of the vibration data was directed toward identifying the most frequently occurring source of the impulses.

Two methods of determining the impulse source were utilized. To illustrate these two methods, a typical impulse event is shown in Fig. 10. The first method was based on comparing the magnitudes of

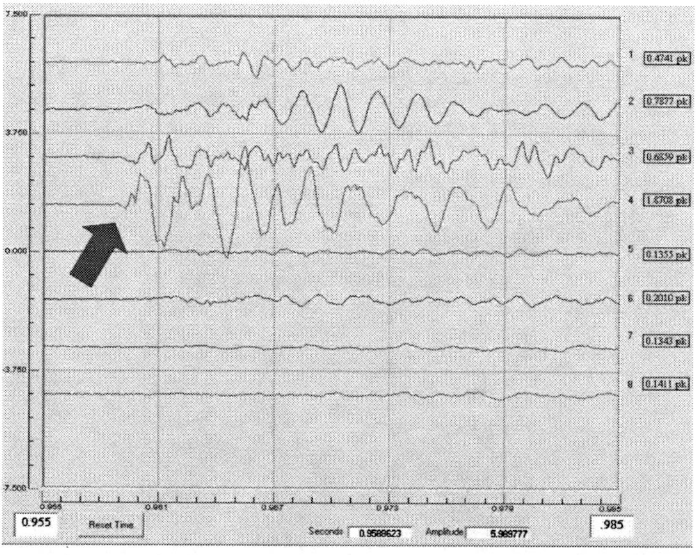

FIG. 10—*Typical impulse event showing time-histories of accelerations.*

vibration measured by each accelerometer. This method had the potential to clearly identify the mullion that was closest to the source of the impulse and appeared to also identify whether the source was at the top or lower horizontal rail. It was based on the premise that the vibration amplitude would dissipate with distance. However, the fact that the sensors were attached to the snap-on covers over the mullions introduced amplification effects that obscured the structurally borne vibrations.

The second method was based on the arrival time of the first wave at each location. Although this was a more tedious method, it provided a more reliable determination of the likely source of each impact. The event shown in Fig. 10 illustrates that, in this case, either method would identify the source of vibration as being closest to Gage 4 (fourth trace from the top), at the top rail adjacent to Mullion 2. The wave front arrived first at Gage 4 and its magnitude of vibration was the largest of the eight sensors.

A tabulation of impulse source locations for 397 events in 2 days revealed that 159 events originated at Mullion 2 and 238 events originated at the other three instrumented mullions. At Mullion 2, 145 of the 159 events (91 %) originated near the top rail. At the other mullions, 140 of the 238 events (59 %) originated closer to the lower rail.

Analysis of Findings and Implementation of Repairs

The cause of the popping noises was identified through visual examination of the connection details and analysis of vibration and deflection measurements. The problem was caused by restrained thermal expansion that progresses gradually from west-to-east as the sun crosses the sky. The concave shape and orientation of the curtain wall combined to exacerbate the condition. If the entire length of the concave wall received the warming effects of the sunshine at the same time, the problem would not have been as severe. As the sun rises, it begins to shine on the west end of the curved wall, causing it to begin expanding. At that point, the remainder of the wall is still in the shade, so the expansion of the west end pushes toward the east. As the sun continues to rise and move westward, the west end of the curved wall gets progressively warmer and increases its expansive push toward the east. The west end has already expanded and is bearing against the restraint at the west end of the wall.

As the thermal expansion builds, the expansive force overcomes the frictional resistance at the connection between the horizontal rails and the vertical mullions. On a flat wall utilizing this type of curtain wall, the horizontal rails are intended to slip easily past the mullions because of the parallel aluminum-to-aluminum contact surfaces. Relatively low load is required to produce the slippage and very little dynamic energy is released. However, in this case, the rails meet the mullion at a slight angle (4° from parallel) producing more friction at the point of contact. More force has to accumulate before the friction is overcome and slippage occurs. The release of the higher bound-up energy level results in a larger dynamic response and corresponding noise in the frame of the curtain wall.

By the time the sun is shining on the entire curved face of the wall, the full thermal load has pushed the system eastward, resulting in a large buildup of force acting on only the east end of the wall. This results in the largest impulses occurring at the east end of the wall at the time when the wall is fully exposed.

The vibration measurements verified that the impulsive vibration events originated at both the top rail and lower rail. This clearly indicated that remedial actions would be required at both levels of the curtain wall. Teflon® shims were recommended to be installed between the bearing surfaces at the connections between the horizontal rail and the vertical mullion, This required loosening the bolts that clamp the two elements together, placing a preformed shim in the gap and retightening (but not overtightening) the bolts by hand. At the top rail connections, there was not enough room to provide access for this process. Therefore, a silicone-based lubricant compatible with existing glazing materials was selected to be injected into the gap between the rail and mullion. These repairs have successfully mitigated the popping noises over the past two years without any supplemental maintenance.

Lessons Learned

The lessons gleaned from this case history and similar projects are summarized below.

- Dynamic data acquisition systems consisting of accelerometers and a dynamic signal analyzer can be used to locate the source of impulsive vibrations produced within a structure.

- The authors have experienced about a 75 % success rate (roughly three out of every four projects) in mitigating disturbing building noises using the methodology described herein.
- It may be prudent to verify the effectiveness of curtain wall connections designed to slip, especially in atypical applications. In this case, an erroneous assumption was made that the connections would slip in the same manner as a typical flat wall and that the orientation of the wall with respect to the sun would have no effect on the system.

References

[1] Berendt, R. D., Winzer, G. E., and Burroughs, C. B., "A Guide to Airborne, Impact, and Structure Borne Noise—Control in Multifamily Dwellings," U.S. Department of Housing and Urban Development, 1967.

Journal of ASTM International, Vol. 4, No. 9
Paper ID JAI100862
Available online at www.astm.org

Matthew Browne,[1] Peter Irwin,[1] Jason Garber,[1] Michael Cicci,[1] Jan Dale,[1] and Russ Parnell[1]

Techniques for Predicting Cladding Design Wind Pressures

ABSTRACT: With advances in wind tunnel methodologies in recent years, the measurement of wind loads on wind tunnel models of buildings and structures can be carried out with a high degree of accuracy. These results lead to more economical and risk-consistent structural designs than do code calculation methods. Five topics related to the prediction of cladding design wind pressures are examined in this paper: (1) wind tunnel testing and analysis methods, (2) integration of wind tunnel data with the full-scale wind climate, (3) evaluation of internal pressures, (4) duration of constant wind pressure for determining the load resistance of glass, and (5) computational fluid dynamics (CFD).

KEYWORDS: cladding design, wind tunnels, wind climate, internal pressures, load duration, CFD

Introduction

Building codes and standards provide minimum loads that buildings and other structures should be designed to resist, which includes wind loads. The strength and direction of the wind varies continuously with time so that, unlike most of the other imposed loads that are static, the wind loads are dynamic. Wind loads are sensitive to the shape of the structure and to its surroundings, and can even be influenced by the movement of the structure itself, if it is sufficiently flexible. The probability of a given level of load occurring is also a function of the building orientation and the joint probability of wind speed and direction at the site. Accordingly, the writers of codes and standards have a very hard task when it comes to wind. Specifying precise and accurate design wind loads for every possible building shape and site environment would result in load provisions so bulky and complex as to be no longer useful to practicing structural designers. Over-simplification results in undue conservatism if all possible values are to be safely contained or to the risk of some cases of under-design if the overall level of conservatism is reduced. Therefore, the writing of good codes and standards involves striking the right balance between accuracy and simplicity. While, typically, the effort to simplify will result in wind load provisions that err on the conservative side for most buildings, they may also result in underestimates that erode some of the factor of safety for some cases.

In recognition that every possible situation cannot reasonably be covered and that conservatism may impose undue cost penalties in certain cases, most modern codes and standards now recommend project-specific wind tunnel tests where accuracy in the wind loads is important for cost and safety reasons and where more complex wind conditions are present, or both. With respect to wind tunnel tests, the American Society of Civil Engineers (ASCE) 7-02 Standard [1] has this to say: "Wind tunnel tests are recommended when the building or other structure under consideration satisfies one or more of the following conditions: (1) Has a shape which differs significantly from a uniform rectangular prism or "box-like" shape; (2) Is flexible with natural frequencies below 1 Hz; (3) Is subject to buffeting by the wake of upwind buildings or other structures; or (4) Is subject to accelerated flow caused by channeling or local topographic features." In fact, most of the generic wind load provisions given in codes and standards are themselves based on wind tunnel results. Full-scale data, which tends to be sparse and building-specific, is used to provide benchmarks for comparison with the wind tunnel data. Therefore, it makes perfect sense to direct the design team back to the original source if the extra cost of the wind tunnel studies can be justified based on the size of the project and the consequences of the wind loads for cost and safety.

To illustrate the typical differences between wind tunnel results and those obtained from the ASCE

Manuscript received November 3, 2006; accepted for publication August 23, 2007; published online September 2007.
[1] Rowan Williams Davies & Irwin Inc., 650 Woodlawn Road West, Guelph, Ontario, N1K 1B8.

TABLE 1—*Comparison of peak suction pressures between wind tunnel and ASCE 7-02.*

	Plan	WT Typical (psf)	WT Range (psf)	ASCE 7-02 Central (psf)	ASCE 7-02 Edge (psf)
Building#1 H=186 ft		-60	-60 to -100	-70 (-93)	-127 (-151)
Building #2 H=250 ft	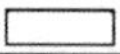	-75	-75 to -130	-90 (-121)	-166 (-197)
Building #3 H=253 ft		-75	-75 to -130	-91 (-122)	-166 (-197)
Building #4 H=318 ft		-80	-80 to -150	-95 (-128)	-174 (-207)
Building #5 H=337 ft	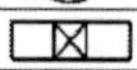	-80	-80 to -150	-96 (-129)	-177 (-210)
Building #6 H=390 ft	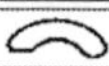	-80	-80 to -140	-99 (-133)	-182 (-216)
Building #7 H=397 ft	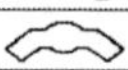	-80	-80 to -175	-100 (-134)	-183 (-217)
Building #8 H=472 ft		-85	-85 to -160	-103 (-139)	-190 (-225)
Building #9 H=539 ft		-85	-85 to -165	-106 (-143)	-195 (-231)
Building #10 H=752 ft		-95	-95 to -165	-114 (-153)	-209 (-248)
Building #11 H=550 ft		-85	-85 to -170	-89	-178

7-02 Standard, Table 1 shows peak suctions for a variety of building shapes in various exposures or surroundings. The values given in brackets represent the suctions on the roof, as defined by the standard. Table 1 illustrates the fact that in most cases the wind tunnel procedure results in lower loads than the ASCE 7-02 analytical method. However, for rectangular buildings in simple surroundings, the comparisons between the wind tunnel and the ASCE 7-02 Standard for the wall suctions are much better, as demonstrated by Building #11 in Table 1. The roof suctions from ASCE 7-02 are not shown for this building.

Wind Tunnel Testing and Analysis Methods

The wind tunnel method has been described in many papers in the literature [2,3]. For wind tunnel methods to be permitted as the basis for design and the results accepted by code officials, the ASCE recommends that wind tunnel tests typically should meet all of the following conditions [1]:

1. The natural atmospheric boundary layer has been modeled to account for the variation of wind speed with height.
2. The relevant macro- (integral) length and micro-length scales of the longitudinal component of atmospheric turbulence (i.e., size of the large- and small-scale turbulence eddies) are modeled to approximately the same scale as that used to model the building or structure.
3. The modeled building or other structure and surrounding structures and topography are geometrically similar to their full-scale counterparts, except that, for low-rise buildings meeting the requirements of Section 6.5.1, tests shall be permitted for the modeled building in a single exposure site as defined in Section 6.5.6.1.
4. The projected area of the modeled building or other structure and surroundings is less than 8 % of the test section cross-sectional area unless correction is made for blockage.
5. Reynolds number effects on pressures and forces are minimized.
6. Response characteristics of the wind tunnel instrumentation are consistent with the required measurements.

Further to the above general requirements, a detailed manual of practice for wind tunnel testing has also been written by the American Society of Civil Engineers [4]. A separate ASCE standard for wind tunnel testing is currently being developed.

Having satisfied these requirements, the data recorded during the wind tunnel tests should be representative of those expected at full-scale and be accepted by building officials.

Wind Tunnel Test Model

The wind tunnel model consists of the study building, along with immediate surrounding buildings. The study building is nowadays usually constructed using rapid prototyping technology (e.g., stereolithography) at a scale in the range of 1:300 to 1:500. In short, the process of model construction using stereolithography for a cladding pressure study consists of three steps:

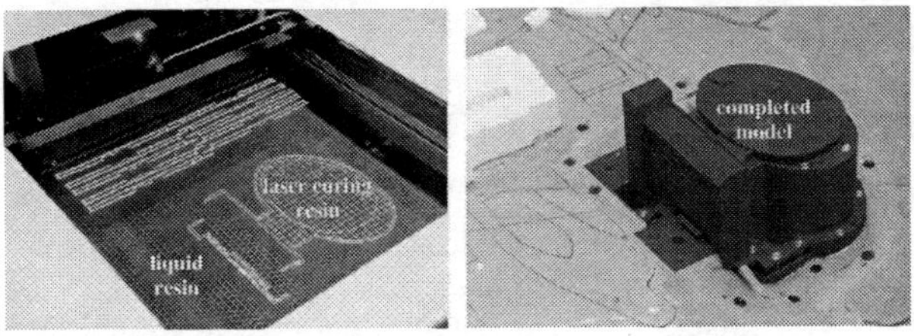

FIG. 1—*Stereolithography method of model construction.*

1. Create a three-dimensional (3-D) solid model of the building geometry;
2. The model is "grown" during a process in which photosensitive resin is cured by a laser in very fine layers (see Fig. 1); and,
3. The model is instrumented with pressure taps. Unlike traditional construction techniques, the stereolithography technology allows complex structures to be modeled accurately and relatively quickly.

Surrounding buildings are modeled within a radius of typically 1200 to 1800 ft. Beyond this radius, the wind tunnel facility simulates the upwind terrain by using a long working section with a roughened floor and specially designed turbulence generators, or spires, at the upwind end (Fig. 2). This working section will generate the proper variation of the mean wind speed and turbulence intensity profiles with height for the wind approaching the modeled surroundings [5]. As the approaching flow interacts with the modeled surroundings, wind flow patterns and turbulence levels similar to those expected at the site are simulated.

The raw output from cladding pressure tests will usually consist of the mean, standard deviation (often referred to as RMS or root-mean-square), and peak positive and peak negative values of the pressure coefficients at each pressure tap. The data are sampled over a time period consistent with the intensity and duration of the design wind event. The data are collected for 36 wind directions at 10° intervals, which is typically sufficient to fully describe the dependence of the wind loads on wind direction. A pressure coefficient C_p is defined as

$$C_p = \frac{p - p_{ref}}{q_{ref}} \tag{1}$$

where p is the local pressure on the model or building surface, and p_{ref} and q_{ref} are, respectively, the reference static pressure and reference dynamic pressure of the wind well above the site at the top of the simulated atmospheric boundary layer. The reference static pressure is the ambient atmospheric pressure in the free stream wind flow above the building. It is read by the static pressure tap on the side of a standard pitot-static probe aligned with the wind in the free stream wind flow. The reference dynamic pressure is the pressure caused by the wind flowing with a positive velocity. The symbol p is used here to represent whichever particular pressure quantity we are concerned with, typically either a peak positive or peak negative value. Peak pressure coefficients may be estimated directly, but it is preferable to sample a

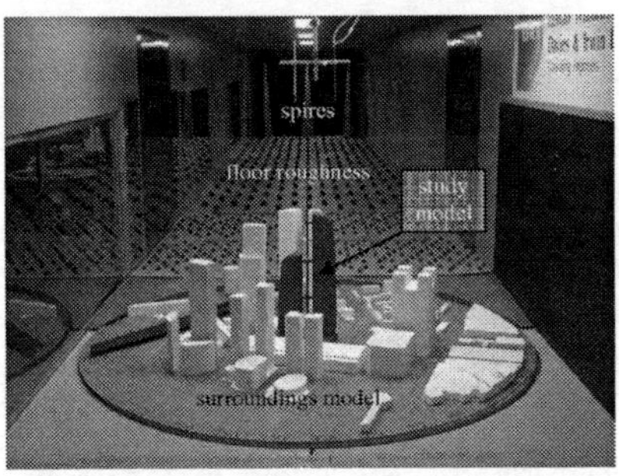

FIG. 2—*Boundary-layer wind tunnel.*

population of measured peak values and then use statistical methods to evaluate the expected peak value during a storm [5].

Integration of Wind Tunnel Data with the Full-scale Wind Climate

The pressure coefficients obtained during the wind tunnel tests at hundreds of locations over the building envelope can then be readily converted to full-scale pressures for any wind speed and direction. Codes and standards, such as the ASCE 7 Standard, specify design wind speeds corresponding to the 50-year return period in nonhurricane wind climates, and somewhat longer for hurricane climates. The return period of an event is defined as the inverse of the annual probability of its occurrence (i.e., a 50-year return period wind speed has a probability of occurrence of 0.02).

As a first approximation, the code-specified wind speed may be applied with the worst-case pressure coefficients (i.e., the highest pressure coefficient for a given location, from all of the wind directions) from the wind tunnel data to provide a "nondirectional" prediction of the wind pressures on the building façade. The reason this is called "nondirectional" in that no consideration is given to the combination of the sensitivity of the pressure coefficients or the full-scale wind statistics to the wind direction.

Such an approach is conservative, as it aligns the design wind speed with the most aerodynamically sensitive direction for every point on the building façade. In reality, different locations will be sensitive to different wind directions, and different wind directions will be more or less likely to be associated with strong winds. The ASCE specifies a directionality factor (Kd) of 0.85 on pressure to reflect this, although the factor is calibrated to be used with load factors that may not necessarily be used for glazing design. Nevertheless, this can be seen as an acknowledgment that the nondirectional approach is conservative.

A wind tunnel study allows the designer to take full account of the joint probability of wind speed and direction, as well as the variation of aerodynamic coefficients with wind direction. This can be accomplished through the use of hourly measurements of wind speed and direction, taken at a nearby anemometer station such as those located at most airports (typically 30 to 50 years of recorded data are readily available for airports located in North America). For sites that are prone to hurricanes or typhoons, wind speeds from a Monte Carlo simulation for the region are typically required since the occurrences of hurricanes at a particular site are too rare to provide meaningful data for a statistical analysis. Various techniques exist to integrate the wind tunnel data with the wind statistics (real or simulated) to provide predicted design pressures corresponding to the design return period. Three common techniques or methods are described in the following paragraphs.

The first approach, which has found use in wind codes in the U.K. and Australia, involves the use of "sector wind speeds." Simply put, wind directions associated with stronger winds are assigned higher wind speeds, and the remaining wind directions have reduced speeds. Therefore, the wind speed for each direction is applied to the wind tunnel data for each direction, and then the maximum pressure is used for each façade location.

This approach is attractive in that it makes intuitive sense to the designer, and is readily adapted in code analytical methods. However, there are limitations in separating out the directionality of the wind speeds from the directionality of the building aerodynamics. The approach is described by Holmes [6], and is used by the authors primarily for situations where it is required to consider specific wind directions to generate wind load patterns for structural design (e.g., on large span roofs). In these situations, the predictions are still calibrated against one of the other techniques mentioned below.

In reality, what the cladding designer requires is the pressure at a given location that corresponds to the desired return period. An analysis method that is finding more common use among wind engineering consultants is sometimes referred to as a "time history approach" [7]. The procedure is relatively straightforward although computationally intensive. From the hourly wind records (i.e., wind speed and direction) the hour-by-hour gradient wind pressures are derived and are then applied directly to the pressure coefficients from the wind tunnel test. The result is a time history of peak full-scale pressures acting at the individual measurements locations of the building covering every hour of the full-scale record. From this record, extreme value techniques can be used to generate the predictions corresponding to the desired return period.

The above approach is relatively new, mainly due to the computational effort that is now easily achieved with the processing speed and storage capacity of a desktop computer. Prior to this, the "up-

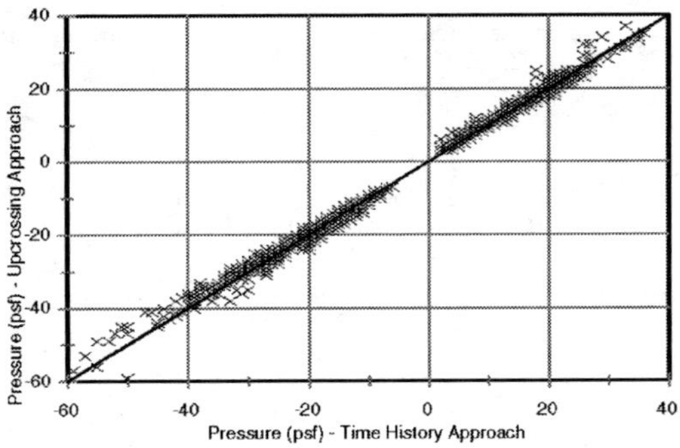

FIG. 3—*Comparison of 50-year return period peak external pressures from upcrossing and time history approach for a building in the Seattle area.*

crossing technique" [8] was an approach that replaced much of the repetitive calculations with analytical expressions. In general, this method involves the determination of the wind speed boundary associated with a given load effect level (i.e., local wind pressure) as a function of wind direction. During the design life of the structure or element, should the wind speed cross the boundary in an upward direction, the wind pressure will then exceed the given level. The associated crossing rate, and subsequently the return period, for a given wind pressure level can be calculated using the boundary information along with the statistical model of the local wind climate. Unlike the time history approach (which uses hourly readings directly), the upcrossing technique requires the hourly wind speed records to be fitted using mathematical expressions. The Weibull fitting technique is commonly applied to the hourly records to generate a representation of the statistical wind climate.

The authors employ both the time history and upcrossing techniques, depending on the situation. A comparison of the predicted peak local exterior pressures using the two techniques is given in Fig. 3 (detailed comparisons are given in Ref. [7]).

The data shown in this figure are based on applying the two analysis methods to the same pressure coefficient data for an example building at approximately 600 measurement locations. If the two methods provided identical results, the points shown in Fig. 3 would fall exactly on the diagonal line. The two techniques give essentially the same result, bearing in mind the uncertainties associated with defining full-scale wind climates.

When a Monte Carlo simulation for a hurricane climate is employed, which generates the equivalent of many thousands of years of data, the time history approach is robust and avoids some of the approximations required in the intermediate step of "fitting" the wind statistics. On the other hand, in areas where the wind statistics are sparse, or where datasets from multiple sites are combined, "fitting" the data for the upcrossing technique affords the opportunity to use some judgment as to whether the statistical model makes physical sense. For example, a short period of record may leave certain wind directions out which would otherwise have data points once more time has passed. Both the time history method and the upcrossing technique typically yield lower results than the nondirectional (or upper bound) approach. Depending on how the wind tunnel data align with the dominant wind directions of the local climate, these "directionality" reductions can be small or large. Since all wind records are subject to some uncertainty, regardless of the length of the record, the authors use lower limits to the directionality reductions (i.e., the ratio of the final prediction to the nondirectional value) that may result from the time history and upcrossing techniques.

Evaluation of Internal Pressures

Internal pressure can be a significant component of the overall net pressure experienced by cladding elements during wind storms. In strong winds, air leakage effects dominate internal pressures. Important sources of air leakage in buildings are window breakage due to airborne debris in a wind storm and open windows, in cases where they are operable. These internal pressures can be on the order of about 10 % to

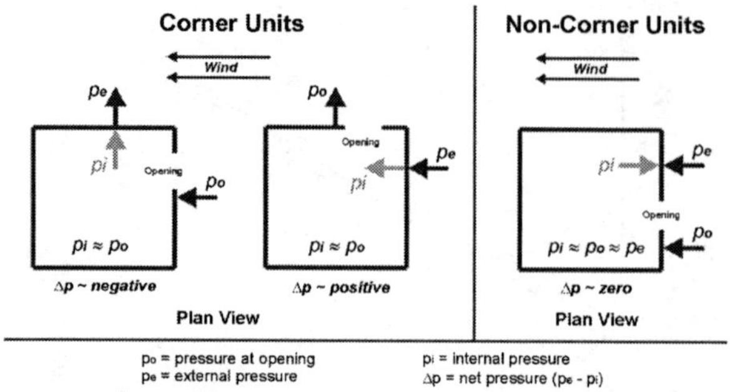

(a) *Corner Units* (b) *Non-Corner Units*

FIG. 4—*Internal pressure cases.*

sometimes even 50 % of the overall pressure exerted on cladding elements. As such, internal pressure allowances are included in recommended cladding design wind loads to account for the increase in pressure when a breach in the building façade occurs.

For corner apartments or units, where orthogonal building faces are exposed to wind, an opening in one façade element could exert high internal pressure to the orthogonal façade element in phase with the high external pressure acting on it. As outlined in Fig. 4(*a*), this could result in very high net overall pressures. In the past, the authors used an in-house approach [9] to incorporate the risk of an opening in the determination of an internal pressure allowance for corner units. More recently, the authors have developed a new methodology for estimating the internal pressures for this case. As outlined in Fig. 4(*b*), in the case of a dominant opening occurring in the façade of an apartment with only a single face exposed to the exterior, the internal pressures could in fact decrease the loading. In such a situation, instead of assuming the internal pressures resulting from the dominant opening, the authors typically consider nominal leakage suggested by the applicable code or standard as the internal pressure.

The methodology developed to estimate internal pressures on a corner region links research considering windborne debris damage and the results of wind tunnel pressure tests. The random and chaotic nature of many of the variables involved in the cause of an opening in the building façade, and the subsequent formation of internal pressures, makes a deterministic solution almost impossible to achieve. As such, a probabilistic approach is used to estimate internal pressures at a level of risk consistent with the design external pressures.

To evaluate the internal pressures resulting from openings in the building envelope, measurements are taken at potential opening locations (i.e., building corners) during wind tunnel testing. These data are analyzed to determine the range of internal pressures that may occur at the selected opening locations, and for a range of probabilities of these openings occurring. The results of research presented on the subject of damage due to windborne debris are used as the basis for estimating the probability of a dominant opening occurring on a building façade based on its geographic region. Also, as impact breakage is more likely to be associated with positive external pressures, differences in relationships between this probability of an opening and the positive and negative internal pressure scenarios are considered. As outlined below, internal pressures are estimated corresponding to a selected reasonable risk level, or return period (usually taken as equal to the level of risk assumed for the external pressures), using the probability of an opening applied to the results of wind tunnel tests. As mentioned earlier, the return period of an event is defined as the inverse of the annual probability of its occurrence (i.e., a 50-year return period pressure has a probability of occurrence of 0.02).

The following is an illustrative example outlining the procedure used to estimate internal pressures from wind tunnel results.

Step 1—Determination of Base Net Pressures

Wind tunnel measurements are used to determine base internal pressure values. This is done by subtracting the external pressure measured at a given pressure tap from a measurement of the instantaneous net

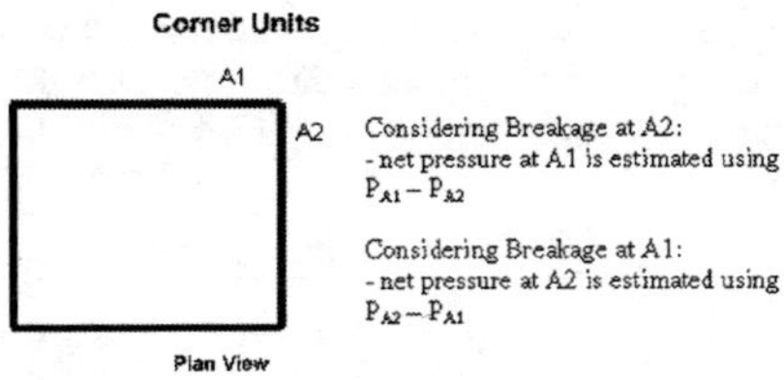

FIG. 5—*Diagram of pressure tap locations and possible net combinations.*

pressure as outlined in Fig. 5. In doing this, four possible breakage scenarios are considered, outlined in the following example. Consider the pressure tap distribution outlined in Fig. 5.

Given: - Corner instantaneous net measurement A1−A2=110 psf, −75 psf
- External pressure at A1=65 psf, −56 psf
- External pressure at A2=41 psf, −65 psf

Considering the pressures at tap A1 (i.e., breakage at A2) the possible combinations are
$65-110=-45$ internal, and
$-56-(-75)=19$ internal.

For tap A2 (i.e., breakage at A1) the net channel is reversed yielding
$41-75=-34$ internal, and
$-65-(-110)=45$ internal.

In cases where a tap has been used in more than one virtual combination, the worst case internal pressure is selected. The resulting internal pressures are then added to the design external pressures to produce base net pressures.

Step 2—Determination of Scale Factor

The probability of a dominant opening (i.e., breakage or open window) occurring is assessed. The overall probability of attaining the internal pressures measured in the tunnel is assessed using the probability of a dominant opening and the probability associated with the wind event return period (i.e., 50-year return period winds have an occurrence probability of 0.02). The product of these values gives the combined probability, or return period ($RP_{Breakage}$), of having breakage during the peak storm event. It should be noted that the basis for a design safety factor in limit-states design is to factor the design load to the "ultimate" limit state. In general, for 50-year return period wind loading, the associated "ultimate" return period ($RP_{Ultimate}$) is taken to be about 500 years. This is a probability step of 0.1, which has been used for all design return periods for the purposes of the internal pressure calculations (i.e., ultimate for a 100-year return period design load is assumed to be 1000 years). The base net pressures, determined in Step 1, are scaled to the ultimate return period from the breakage return period as follows:

$$[0.5 * \log_{10}(RP_{Ultimate}) + 0.15]/[0.5 * \log_{10}(RP_{Breakage}) + 0.15]$$

This equation was empirically developed to produce similar factors as those presented in Table C6-3 from the ASCE 7-02 Standard. For example, when $RP_{Ultimate}=500$ and $RP_{Breakage}=50$, this scale factor equals 1.5 on pressure.

Step 3—Corner Zone Internal Pressure Selection

The external pressures are combined with a selected internal pressure value to produce a design net pressure. The ratio of the ultimate base net pressures to these design net pressures is calculated. The selection of this internal pressure value is based on obtaining an ultimate-to-design ratio of less than the safety factor to be used for the design for a reasonable percentage of the pressure taps. The remaining pressure taps, having an ultimate-to-design ratio greater than the design safety factor, are assessed on a case-by-case basis.

Note that internal pressures are included to prevent subsequent failure due to the increase in pressure resulting from a dominant opening occurring in the façade. Increasing the design pressure of a cladding element will not necessarily increase the impact resistance. As such, breakage is still likely to occur due to debris impact in the event of strong winds. The recommended internal pressure allowances are assessed to prevent subsequent failure due to the increase in pressure in the event of a façade breach. To reduce the chance of breakage due to windborne debris, impact-resistant glazing should be used for the entire cladding system.

Duration of Constant Wind Pressure for Determining the Load Resistance of Glass

For glazing design it is desirable to know the duration for which the loads are applied, because the strength of glass, although not highly sensitive to the load duration, does have some dependence on it. Typically, the strength is proportional to $t^{-1/16}$, where t is the load duration. According to this relationship, the glass strength would, for example, go down by about 7 % if the load duration were increased by a factor of three. In the wind tunnel the typical duration of individual peak loading events corresponds to about half to one second at full scale. However, in a storm of several hours the same loading level may be reached a number of times. Dalgliesh [10] has considered this point and discusses the final failure of glass as being due to accumulated damage after being heavily loaded. Since glass takes a significant time to "heal" from damage after being heavily loaded, the accumulated load duration for a storm could be considerably longer than a half to one second. Added to this is the accumulated damage effect of all of the loads that are slightly less than the maximum values. With these considerations in mind, it has traditionally been recommended that the load duration for glazing be taken as 60 seconds. This duration has also been used, until recently, in the American Society for Testing Materials (ASTM) E 1300 Standard for the specification of glass strength.

The design of glazing elements for buildings based on the ASTM E 1300 Standard involves selecting a specific thickness of glass lite so that the factored load resistance (LR) exceeds the specified load. The LR value is obtained from the ASTM E 1300 Standard, while the specified load is determined from relevant building codes, typically the ASCE 7 Standard.

This section will outline the inconsistency of LR values as prescribed by the ASTM E 1300 Standard from the 98 edition [11] to the 02 edition [12], when using the ASCE 7 Standard to specify design loads. The specific interest is short-duration loads, which are applicable for the design of glazing components in the cladding systems of buildings.

The specific inconsistency between the ASCE and ASTM Standards is that there is a specified load duration for applied loads in the ASTM Standard, while the ASCE Standard does not have a duration associated with the specified design loads. Section 3.2.8.4 of ASTM E 1300-98 refers to a nonfactored load (NFL) as a "sixty second duration uniform load associated with a probability of breakage of 8 lites per thousand…". Section 3.2.7.4 of ASTM E 1300-02 refers to a nonfactored load (NFL) as a "three second duration uniform load associated with a probability of breakage of 8 lites per thousand…". The nature of damage accumulation on glass is such that there is a significant increase in the capacity of a glass pane when the load duration is reduced from 60 to 3 seconds (for more information see Ref. [13]).

There is no load duration associated with the specified loads from the ASCE Standard. ASCE 7-93 [14] and earlier editions specify a fastest-mile wind speed, with no reference to gust duration. ASCE 7-95 and newer editions specify a three-second gust wind speed. It could be implied that the load duration for a three-second gust design wind speed is three seconds. It is important to note that the implied duration is not specified by the ASCE 7 Standard, and designers will not necessarily account for this or understand its importance for the selection of glass thickness.

Calculating the glass design loads and design thicknesses for an example building in Houston can be examined to illustrate the impact of the change in load duration by the ASTM E 1300 Standard. The design load for cladding elements in Zone 4 (i.e., central wall area) determined by various editions of the ASCE 7 Standard was used to select a thickness of single-pane annealed monolithic glazing based on the appropriate ASTM Standard. Table 2 shows the design loads and glass thicknesses.

Despite the change in the ASCE 7 wind speed contour map, as well as the specified wind speed type, there is no change in glass thickness based on the ASTM E 1300-98 Standard when the ASCE 7-93 and ASCE 7-98 [15] design loads are considered. It should be noted that the LR/Design Load value of 0.99 should be acceptable as the next greater thickness increment (12 mm) would result in a LR/Design Load of 1.69.

The change in glass thickness between the ASCE 7-98 and ASCE 7-02 case can be attributed to the

TABLE 2—*Glass thickness comparisons.*

ASCE 7 Edition	7-93 [14]	7-98 [15]	7-02 [1]
Design Wind Speed	93 mph	110 mph	110 mph
Wind Speed Averaging Time	fastest mile	3-second gust	3-second gust
ASTM E 1300 Edition	E 1300-98 [11]	E 1300-98 [11]	E 1300-02 [12]
Load Duration	60 second	60 second	3 second
Design Load from ASCE 7 (Zone 4)	53.7 psf	51.2 psf	51.2 psf
Glass Plate Width × Length	39.4 in. × 78.7 in. (1000 mm × 2000 mm)	39.4 in. × 78.7 in. (1000 mm × 2000 mm)	39.4 in. × 78.7 in. (1000 mm × 2000 mm)
Glass Thickness (Zone 4)	0.39 in.(10 mm)	0.39 in. (10 mm)	0.31 in. (8 mm)
LR/Design Load (Zone 4)	0.99	1.04	1.225

updated NFL tables and the ASTM E 1300 Standard. Therefore, the same design load can be resisted by a thinner lite of glass in 2002 than in 1998. If this is not due to increased performance from manufacturing improvements, then it is likely that glazing elements based on the ASTM E 1300-02 Standard have a higher risk of failure than those based on the ASTM E 1300-98 Standard. The higher risk of failure using the ASTM E 1300-02 Standard applies to all cases, not just the example shown in this paper.

Computational Fluid Dynamics (CFD)

The field of computational fluid dynamics (CFD) has advanced considerably in scope and accuracy over the past few decades, to the position where accurate predictions of flow speeds and pressures can be predicted in situations where turbulence is not dominant. All CFD methods make simplifying assumptions to deal with turbulence. Even the best "large eddy simulations" (LES), which seek to represent the larger time-dependent turbulent gusts, include some simplifying assumptions for the smaller gusts in order to reduce the computational overheads. These assumptions limit the usefulness and accuracy of CFD for building aerodynamics, especially for the determination of wind loads where high accuracy is needed, because turbulence is the dominant influence on the flow. The largest and fastest CFD facility available today would take thousands of years to calculate one hour of wind at full scale to the same detail provided by the typical wind tunnel simulation. However, because of the way that the model scale factors compress time, the wind tunnel achieves the same hour at full scale in only about 36 seconds of "tunnel time." The wind tunnel may be regarded as an analogue computer that calculates all parts of the flow simultaneously at a grid size of around 3.3×10^{-5} ft (10^{-5} m), corresponding to around 0.12 in. (3 mm) at full scale for a typical linear scale factor of 1:300. While it is always necessary to resolve the whole flow for a CFD simulation, the wind tunnel need only sample the desired information from the analogue simulation. Thus, it is fair to say that the wind tunnel still provides better information faster and more cheaply than CFD. If anything, the improved efficiencies and capabilities of wind tunnels that have come from advances in computing power, electronic instrumentation, and computer-based rapid prototyping have actually widened the gap between the capabilities of wind tunnel testing and CFD in recent years for wind engineering applications.

To illustrate the gap that currently exists between CFD and the wind tunnel, mean pressure coefficients for L- and U-shaped buildings are shown in Fig. 6 [16]. The researchers found good agreement between experimental results for upwind Cp values. However, large discrepancies were found for Cp values in the wake regions as shown in Fig. 6. The typical difference in Cp values is around 0.2, which corresponds to 30–35 % lower wind loads from the CFD simulation compared to the wind tunnel tests. Note that these coefficients are only the mean component of the wind loading and, therefore, do not contain the fluctuating component, which is important for obtaining the local peak pressures used in design. The discrepancies shown in this figure are representative of the current state of CFD for predicting detailed pressures on buildings (see Ref. [17] for another example).

From the research to date, it is clear that CFD is not appropriate for determining detailed design wind loads on buildings at present; however, it does have some useful applications within wind engineering [18]. For example, CFD provides good visualization of complex wind flows. Another advantage of CFD is that comparison of a variety of building configurations can be made relatively quickly and cost effectively.

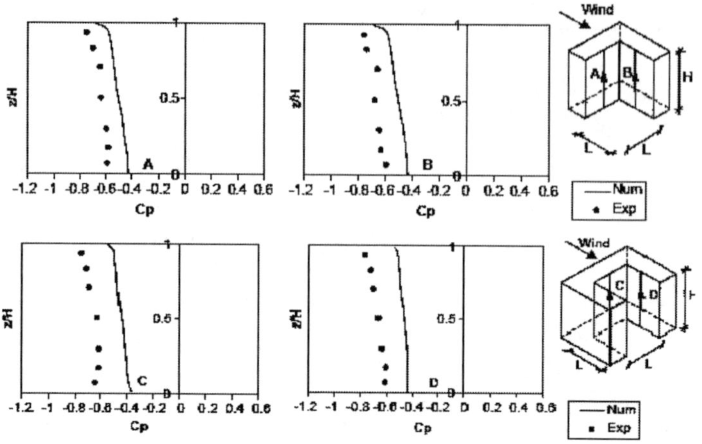

FIG. 6—*Pressure coefficients for L- and U-shaped buildings for normal incident angle of 180° (after Gomes et al. [16]).*

CFD has also proved a useful tool to augment wind tunnel measurements of flow over hilly terrain or other complex topography [19], pedestrian level winds, thunderstorms, and tornados [20]. The following paragraphs highlight two of these applications: pedestrian level winds (PLW) and thunderstorms.

Pedestrian Level Winds (PLW)

Pedestrian level wind simulation is perhaps the most promising area for the use of CFD in wind engineering at present [21]. This area is also one where CFD and wind tunnels can be used to complement each other. An initial CFD study, while not precise in the same way the wind tunnel is, can be used to improve a design and indicate where possible problems may occur. With this information velocity probes can be placed on a wind tunnel model in the areas identified by CFD as being of most interest. Figure 7 shows one such study, where CFD simulation identifies hot spots prior to wind tunnel tests.

Thunderstorm Winds

Thunderstorms are highly localized events that can create high wind speeds close to the ground, in the lower 300 to 650 ft, that typically last only a few minutes. The simulation of such winds in a wind tunnel is still an area of research, with existing techniques being focused on the synoptic winds generated by large-scale cyclonic storms. Synoptic winds are caused by the interaction of large pressure systems in the lower atmosphere. Selvam and Holmes [22], Woods et al. [23], and Hangan et al. [24] have carried out numerical and laboratory studies of impinging jets representing thunderstorms.

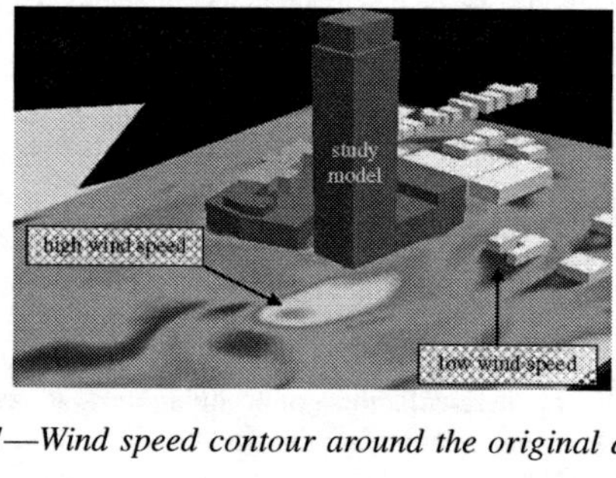

FIG. 7—*Wind speed contour around the original design.*

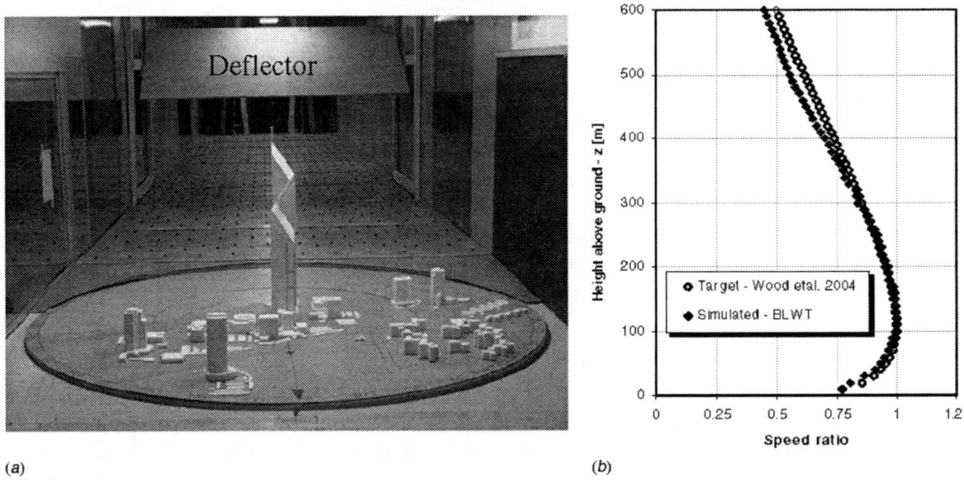

FIG. 8—*(a) BLWT setup to simulate thunderstorm profile; (b) comparison of simulated and target Wood et al. [23] profiles.*

The use of CFD for thunderstorm simulation studies provides significant benefits. For instance, Wood et al.'s thunderstorm profiles have been used by Rowan Williams Davies & Irwin Inc. (RWDI) in setting a special test setup in their 8 ft by 6.5 ft (2.4 m by 2.0 m) wind tunnel in which a low-level jet was created by use of a deflector plate upwind of the study building model as shown in Fig. 7. Also shown is a comparison of the simulated profile in the wind tunnel, with the target from Ref. [23].

It is important to be aware of the limitations of CFD methods in solving practical wind engineering problems. Extensive validation work is required to ensure that a CFD model is applicable for a particular application. At the present stage of their development, CFD techniques are best used to guide preliminary design and as a precursor to wind tunnel studies (Fig. 8).

Concluding Remarks

Wind loads on a building are sensitive to a number of factors, including the wind speed and direction approaching the site, building height and shape, and the local influence of nearby buildings on the wind flow patterns. Wind tunnel methods, which consider all of these factors, determine the wind loading on a structure with increased precision, which leads to more economical and risk consistent structural designs than do code calculation methods. To obtain wind loads suitable for the design of cladding elements, the pressure data obtained from the wind tunnel should be combined with meteorological data, and then augmented with an appropriate internal pressure allowance.

For glazing design, it is important to consider the duration for which the peak pressures should be applied over, which has traditionally been 60 seconds. The change from fastest-mile to three-second gust wind speeds that took place in 1995 in the ASCE-7 Standard did not result in an appreciable change in wind loads for most cases, while the change from 60 to 3-second gust duration in the ASTM E 1300 Standard between the 1998 and 2002 editions resulted in a reduction in the required thickness of glass for the same design load. This resulted in a significant increase in risk of glass failure.

Nowadays a question that comes up frequently is whether CFD has a role to play in determining design wind loads. Due to current limitations in CFD, and the highly turbulent nature of airflows in wind engineering applications, the answer is generally no. Although there are useful applications for CFD within wind engineering, considerable further development and verification will be needed to make CFD a practical tool similar to the wind tunnel.

References

[1] ASCE 7-02 Standard, "*Minimum Design Loads for Buildings and Other Structures*," American Society of Civil Engineers, 2002.

[2] Williams, C. J., Conley, G. J., and Kilpatrick, J., "The Use of Wind Tunnels to Assist in Cladding Design for Buildings," *Performance of Exterior Building Walls, ASTM STP 1422*, P. G. Johnson, Ed., ASTM International, West Conshohocken, PA, 2002.

[3] Irwin, P. A., "The Design of Spires for Wind Simulation," *J. Wind. Eng. Ind. Aerodyn.*, Vol. 7, 1981, pp. 361–366.

[4] ASCE, "Wind Tunnel Studies of Buildings and Structures," ASCE Manuals and Reports on Engineering Practice No. 67, American Society of Civil Engineers, New York, NY, 1999.

[5] Irwin, P. A., "Pressure Model Techniques for Cladding Loads," *J. Wind. Eng. Ind. Aerodyn.*, Vol. 29, 1988, pp. 69–78.

[6] Holmes, J. D., "Reduction Factors for Wind Direction for use in Codes and Standards," Presented at the Symposium on Designing with the Wind, Nantes, France, June 1981.

[7] Gamble, S. L., Miltenburg, R. J., Cicci, M. D., and Accardo, M., "Prediction of Local Exterior Wind Pressures from Wind Tunnel Studies using a Time History Analysis Approach," *Proceedings of the Americas Conference on Wind Engineering*, Clemson, South Carolina, 2001.

[8] Lepage, M. F. and Irwin, P. A., "A Technique for Combining Historical Wind Data with Wind Tunnel Tests to Predict Extreme Wind Loads," *Proceedings of the 5th US National Conference on Wind Engineering*, Lubbock, November 6–8, 1985.

[9] Irwin, P. A. and Sifton, V. L., "Risk Considerations for Internal Pressures," *J. Wind. Eng. Ind. Aerodyn.*, Vols. 77 and 78, 1998, pp. 712–723.

[10] Dalgliesh, W. A., "Assessment of Wind Loads for Glazing Design," *Proceedings of the IAHR/IUTA Symposium*, Karlsouke, Germany, September 3–6, 1979, pp. 696–708 (Springer-Verlag 1980).

[11] ASTM Standard E 1300-98, "Standard Practice for Determining Load Resistance of Glass in Buildings," *Annual Book of ASTM Standards*, ASTM International, West Conshohocken, PA, 1999.

[12] ASTM Standerd E 1300-02, "Standard Practice for Determining Load Resistance of Glass in Buildings," *Annual Book of ASTM Standards*, ASTM International, West Conshohocken, PA, 2002.

[13] Calderone, I. and Melbourne, W. H., "The Behavior of Glass Under Wind Loading," *J. Wind. Eng. Ind. Aerodyn.*, Vol. 48, 1993, pp. 81–94.

[14] ASCE 7-93 Standard, "*Minimum Design Loads for Buildings and Other Structures*," American Society of Civil Engineers, 1993.

[15] ASCE 7-98 Standard, "*Minimum Design Loads for Buildings and Other Structures*," American Society of Civil Engineers, 1998.

[16] Gomes, M. G., Rodrigues, A. M., and Mendes, P., "Experimental and Numerical Study of Wind Pressures on Irregular-plan Shapes," *J. Wind. Eng. Ind. Aerodyn.*, Vol. 93, 2005, pp. 741–756.

[17] Fu, C. L., Lee, S. M., and Cheng, C. M., "Validation of CFD Simulations on the Wind Loads for Tall Buildings' Preliminary Design," *Proceedings of the 4th International Symposium on Computational Wind Engineering*, Yokohama, Japan, July 16–19, 2006.

[18] Bitsuamlak, G. T., "Application of Computational Wind Engineering: A Practical Perspective," *3rd National Conference on Wind Engineering—2006*, Indian Society for Wind Engineering, India, January 5–7, 2006.

[19] Bitsuamlak, G. T., Stathopoulos, T., and Bedard, C., "Numerical Evaluation of Wind Flow Over Complex Terrain: Review," *J. Aerosp. Eng.*, Vol. 17, No. 4, 2004, pp. 135–145.

[20] Selvam, R. P. and Millet, P. C., "Computer Modeling of Tornado Forces on Buildings," *Wind Struct.*, Vol. 6, No. 3, 2003, pp. 209–220.

[21] Wright, N., "Appropriate Use of Computational Wind Engineering," *Proceedings of the 2004 Structures Congress*, Nashville, TN, 2004.

[22] Selvam, R. P. and Holmes, J. D., "Numerical Simulation of Thunderstorm Downdrafts," *J. Wind. Eng. Ind. Aerodyn.*, Vols. 41–44, 1992, pp. 2717–2825.

[23] Wood, G. S., Kwok, K. C.S., Motternam, N. A., and Fletcher, D. F., "Physical and Numerical Modeling of Thunderstorm Downbursts," *J. Wind. Eng. Ind. Aerodyn.*, Vol. 89, 2001, pp. 535–552.

[24] Hangan, H., Kim, J. D., and Xu, Z., "The Simulation of Downbursts and Its Challenges," *Proceedings of the 2004 Structures Congress*, Nashville, TN, 2004.

INVESTIGATION, REHABILITATION AND DETAILING CASE STUDIES

Journal of ASTM International, Vol. 5, No. 4
Paper ID JAI100811
Available online at www.astm.org

Mark K. Schmidt[1] *and Jill H. Murphy*[2]

Erroneous Design Assumptions and Construction Defects in a Thin Stone Facade System

ABSTRACT: During a routine close-up facade inspection of the exterior walls of a 15-year-old, 31-story office building, a loose 30-mm thick granite panel was discovered. Initially, it was believed that this movement was due to the location of the anchorages relative to the panel geometry. A more thorough examination of the panel was performed by inspecting the panel attachment from the interior of the building. This inspection revealed that one of the two epoxied wind load anchors had failed. An extensive investigation of the granite facade panels, including additional close-up exterior and interior inspections, structural analyses, and testing, was undertaken. Interior examinations revealed construction defects in the epoxied anchors. In addition, the structural analyses indicated a flaw in the original design that affected certain panels in high wind load zones. Repairs were designed to address the design and construction deficiencies. Supplemental panel anchorages were installed from the exterior of the building, thereby minimizing disruptions to building occupants. While the implemented panel repairs can be identified at close range, they are imperceptible from publicly accessible vantage points.

KEYWORDS: epoxied pins, granite panels, anchorage design, construction defects

Background

The subject 31-story reinforced concrete high-rise office structure was constructed circa 1989 and is clad primarily with nominally 30-mm thick flame-finished granite panels and an aluminum and glass curtain wall. The granite panels vary in size depending upon their location on the building. Of particular interest for this paper are the 300-mm and 1500-mm wide granite panels supported by a combination of kerf angles and embedded, epoxied pins (threaded rods). Panel-to-panel joints are filled with silicone sealant and panel-to-window joints are filled with 13-mm wide EPDM (ethylene propylene diene monomer) gaskets.

Typically, five pins at nominally 300 mm on center are used to resist wind loads for 1500-mm wide panels and two pins are used for 300-mm wide panels. Each bent threaded rod is installed into an approximately 19-mm deep hole in the panel at a slight upward angle. The horizontal portion of each threaded rod is then connected to an aluminum strongback by means of a nut and washer. Two different types of strongbacks are used for connecting to the structural framing. The first type found behind the 1500-mm wide granite panels is a channel-shaped aluminum extrusion attached to the structural framing at either end as shown in Fig. 1. Five approximately evenly spaced slotted holes are located along the center of the channel web. These holes allow for the attachment of the epoxied pins. The strongback located at the back of the 300-mm wide granite panels is slightly different in construction. Instead of using an extrusion, a formed aluminum hanger is used. The ends of the formed hanger have slots that hook over a round stainless steel rod, which in turn is connected to the structural framing as shown in Fig. 2. Two epoxied pins are used to attach the hanger to the granite panel.

During a required close-up facade examination in 2001, a 300-mm by 2100-mm panel at the seventh floor was found to have minor movement at the top. The connection in question was the epoxied pin connection. Throughout the inspection it was noted that typically the granite panels did have some movement at the top, but to a lesser degree than the panel at the seventh floor. It was recommended that the panel at the seventh floor be monitored and reexamined during the next overall facade inspection.

In November of 2003 as part of a required follow-up inspection, the granite panel located at the seventh floor was reexamined from the interior to determine if its noted movement from the exterior was

Manuscript received September 21, 2006; accepted for publication March 21, 2008; published online April 2008.

[1] Unit Manager and Senior Consultant, Wiss, Janney, Elstner Associates, Inc., 330 Pfingsten Road, Northbrook, Illinois 60062.
[2] Project Engineer, William A. Kibbe & Associates, Inc., 1475 South Washington Avenue, Saginaw, Michigan 48601.

FIG. 1—*Channel-shaped aluminum extrusion attached to back of 1500-mm wide granite panel.*

indicative of a pressing structural issue. To further examine this granite panel, an interior opening was created by removing the interior finishes and then cutting through the sheet metal barrier (see Fig. 3). The interior examination of the panel revealed that one of the pins securing the panel had failed and could easily be removed by hand and the remaining pin was loose (see Fig. 4). It was also noted that stainless steel threaded rods had been used for the pins, but the washers and nuts were carbon steel. Surface corrosion was observed on both the washer and nut of the failed pin and on the loose pin; however, no significant section loss was noted. Immediately following the interior inspection, the panel was temporarily secured to the building.

Investigation

Following the discovery of the failed epoxied pin a comprehensive investigative program was developed. An exterior inspection was performed followed by a structural analysis of the panels and anchorage system. Upon completion of this phase, interior inspections and testing were performed on the panel anchorages.

FIG. 2—*Typical formed aluminum strongback attached to back of 300-mm wide granite panel.*

FIG. 3—*Initial interior inspection opening at seventh floor granite panel.*

Exterior Inspection

Four exterior swing stage inspection drops were performed on the building to determine if the loose panel found at the seventh floor was an anomaly or if loose panels were common throughout the building facade. One drop per elevation was selected with drops varying from the corner of the building to the middle of the building. The purpose was to inspect a variety of panel types and locations. From available drawings, it was determined that there were three types of granite panels on the facade with epoxied pins that

FIG. 4—*Failed epoxied pin shown in reflection mirror.*

FIG. 5—*Power cup assembly for testing relative movement of granite panel.*

required inspection. Two types measuring 1500 mm by 2100 mm and 300 mm by 2100 mm were located at the vision glass level, and a third type located at the spandrel level measured 1500 mm by 1500 mm. Most of the remaining facade panels were attached by means of kerf anchors which did not pose a concern with respect to failure of the epoxied pins.

To determine a qualitative level of movement for each of the three panel types, a test protocol was derived. The first step was to secure the suspended scaffold to the building by using power cups attached to the insulating glass units. Once the scaffold was secure, a power cup specially suited for use on flame-finished granite was attached to the granite panel at the location of the epoxied pins. A handle was attached to the back of the power cup which allowed for pulling on the panel as shown in Fig. 5. In the case of the 1500-mm by 2100-mm panels and the 300-mm by 2100-mm panels, the epoxied pins were located approximately 600 mm down from the top of the panel. For the 1500-mm by 1500-mm spandrel panels, the epoxied pins were located approximately 75 mm above the bottom of the panel. The relative movement of the epoxied pin connections was qualitatively determined for each tested panel. During the inspection drops, sealant and gaskets were also removed along the edges of the panels to verify the locations of the epoxied pins.

Based upon an overall summary of the exterior inspection drop data, it was initially determined that the loose panel condition was related to the hanger-and-rod connection found at the back of the 300-mm by 2100-mm panels. These panels tended to have the most movement in comparison to the other types of panels by virtue of the tolerances associated with the hanger-and-rod connection. Most of the observed movement could be traced to significant play between the slots in the hanger and the stainless steel rod, and minor play between the rod and the structural framing. The 1500-mm by 2100-mm and 1500-mm by 1500-mm panels utilize the aluminum channel-shaped extrusion, which has a stiffer connection to the structural framing.

While the exterior survey did not conclusively identify panels with potentially failed epoxied pins, the various types of panel connections to the structural framing were verified.

Analysis

Prior to further inspection and testing, structural analyses of the three panel types were performed. Initially, an analysis of the 1500-mm by 2100-mm panel with five epoxied pins was completed. In order to account for the relative stiffnesses of the granite panel and the aluminum strongback, a finite element model was created. This model was able to incorporate the prying action that occurred between the relatively stiff granite panel and the more flexible strongback when loaded by wind. While reviewing original design calculations provided to us by the building management, it was found that the granite panel was modeled with the epoxied pins as fixed nodes. No account for prying action had been made. Because of this assumption in the original design, the forces in the two outside epoxied pins were found to be over three times the forces calculated by the original designer. Similar analyses of the 300-mm by 2100-mm and the 1500-mm by 1500-mm panels were also performed.

For the original design of this building, a wind tunnel study was commissioned by the Architect/ Engineer of Record. The wind tunnel study provided wind pressures ranging from 0.96 kPa to 2.3 kPa. Typically, the first six floors were assigned a wind pressure of 0.96 kPa while the remaining floors were assigned a wind pressure of 1.7 kPa with the upper corners of the building being assigned a 2.3 kPa wind pressure. The wind tunnel study wind pressures, which were in excess of the original building code requirements, were utilized when performing the analyses.

Concurrent with the analysis phase, original construction documents and photos pertaining to the installation of the granite panels were reviewed. Determination of possible construction and installation patterns were of interest. From photographs taken during the construction, it appeared that the panels were installed in three-story phases beginning at the fifth floor, which is where the granite panel and vision glass curtain wall begins. This review of documentation proved useful during subsequent interior inspection and testing of the epoxied pins.

Interior Inspection and Testing

Based on the calculated epoxied pin design forces, a statistically-based testing protocol for the epoxied pins was developed in order to determine if the failed pin observed at the seventh floor panel was indicative of a widespread problem. In an attempt to satisfy statistical qualifications, the tested panels were generally randomly selected using a number generator. Panels located around columns or other structural framing interferences were not included in the test sample. Panels were also excluded by building management based on restricted access into some occupied spaces where testing would be intrusive to building tenants.

Interior inspection openings similar to the original inspection opening made at the seventh floor were created at selected panel locations. The inspection openings were made large enough to allow for sufficient inspection of the epoxied pins; however, opening sizes were also kept as small as possible to minimize the amount of repair work required to restore the interior finishes. All inspection and testing work was scheduled such that the openings were created after business hours on Friday and were repaired prior to the start of work on Monday morning. This procedure avoided any unnecessary disruptions to the building occupants.

Once the inspection openings were created, a visual survey of the general conditions of the epoxied pins was conducted. Typically, the as-built conditions matched the original drawings. At some locations, minor surface corrosion, rarely covering the entire element, was noted on the carbon steel nut and washer. The observed conditions did not warrant widespread replacement of the nuts and washers.

A notable finding at the interior inspection openings was the color of epoxy in which the pins were set. Panels were found to either have atypical pink-colored epoxy or typical tan-colored epoxy. Selected panels on the fifth through seventh floors had different colored epoxy than panels observed elsewhere.

Because visual inspection alone did not provide sufficient information regarding the integrity of the epoxied pins, various supplemental evaluation techniques were explored. Common nondestructive testing techniques were dismissed as lacking the precision needed to identify anchorage deficiencies. Instead, a statistically-based proof load test procedure was developed. The pins were tested to a proof load, which corresponded to a multiple of the design forces for the 0.96 kPa and 1.7 kPa wind load zones. The proof load magnitude needed to be high enough to provide the desired degree of confidence in capacity given a reasonable number of tests and low enough to preclude failure of properly installed pins. Since proof testing does not provide a direct estimate of variation, a coefficient of variation had to be assumed for the population of anchors. Detailed derivation of the statistical methodology used is beyond the scope of this paper. Overall, this methodology was utilized to determine how many consecutive successful proof load tests were needed to conclude with a certain confidence level that close to 100% of the population would have an ultimate capacity greater than the factored design load.

The pass/fail proof load test was conducted concurrently with the visual inspection of the epoxied pins. Custom test apparatus was fabricated for use in the field. The width of the test apparatus was adjustable to account for various conditions encountered in the field. The U-shaped test frame was placed so that its legs were bearing on the back side of the granite panel as shown in Fig. 6. A threaded rod with a coupler nut

FIG. 6—*U-shaped test frame used for proof load tests.*

was attached directly to the epoxied pin and was placed through the center of the U-shaped frame. The existing carbon steel nut and washer were removed so that there was adequate pin length for attachment of the coupler nut. At the opposite end of the threaded rod, an alignment washer, a thrust bearing, a load cell, and a nut were placed against the frame. The nut was tightened against the load cell bearing on the frame, which loaded the epoxied pin. Output was displayed directly in pounds. Upon completion of the test, a new stainless steel nut and washer were installed in place of the existing carbon steel nut and washer.

The results of the pass/fail proof load test provided quantitative information regarding the integrity of epoxied pins. All of the epoxied pins tested passed except for pins from the 1500-mm by 2100-mm panels located at the fifth, sixth, and seventh floors.

The results of the pass/fail proof load test were used to determine which types of panels in which wind load zones required repairs. To make this determination, a sufficient number of panels had to be tested in each area of interest (e.g., three-story construction phases) to satisfy the statistical requirements.

Findings

The results of the proof load testing indicated that the epoxied pins in both the 1500-mm by 1500-mm and 300-mm by 2100-mm panels are adequate to resist calculated forces in all wind load zones of the building. While epoxied pin failures did occur during proof load testing of the 1500-mm by 2100-mm panels on the fifth through seventh floors and these results did correlate with observations of different colored epoxy and

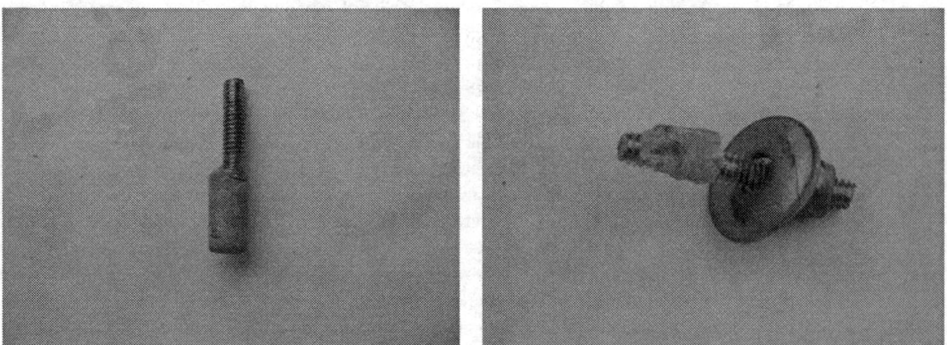

(*a*) (*b*)

FIG. 7—*Examples of failed anchors.*

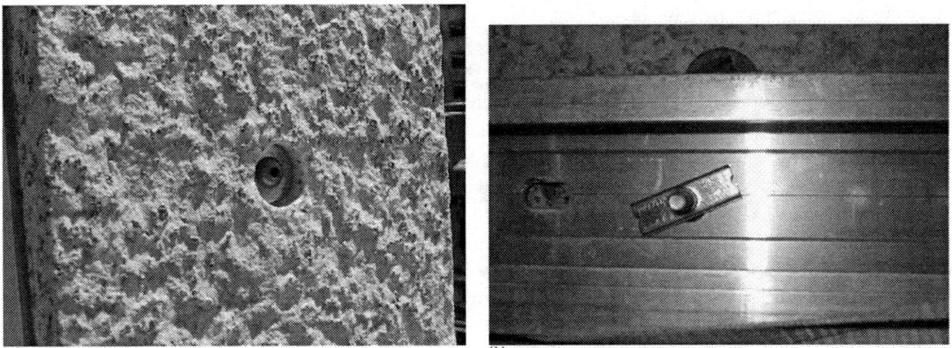

(a) (b)

FIG. 8—*Exterior and interior view of supplemental repair anchor.*

construction phasing, only the 1500-mm by 2100-mm panels on the seventh floor required repair. The similarly sized panels on fifth and sixth floors have a significantly lower design wind load (e.g., 0.96 kPa versus 1.7 kPa). Examples of failed anchors are shown in Fig. 7. The magnitude of the selected proof load was lower than the epoxied pin design loads for the 1500-mm by 2100-mm panels in the highest wind load zone (2.3 kPa). Use of a higher proof load for this zone was considered, but this notion was dismissed. A sufficient proof load for this zone would have approached the epoxied pin capacity determined from mock-ups. As a result, repairs to the 1500-mm by 2100-mm panels in the 2.3 kPa wind load zone were also undertaken.

Repairs

Following the determination of which panels required repair, possible repair details were developed. The repairs had to adequately withstand the wind pressures presented in the wind tunnel study. The repairs needed to be installed from the exterior of the building to minimize disruption to the building occupants. Finally, once installed, the repairs needed to be inconspicuous as viewed from street level and adjacent structures.

A supplemental repair anchor comprised of a stainless steel toggle bolt with a 9.5-mm diameter stainless steel cap screw and washer was developed. Prior to the installation of the repair, several mock-

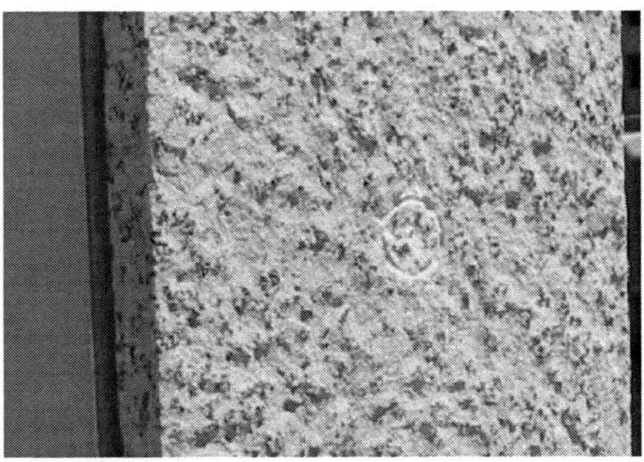

FIG. 9—*Example of repaired granite panel.*

ups were tested in the structural laboratory. The repairs were installed in granite taken from attic stock provided by the building. The results of the laboratory tests correlated well with calculated pullout loads.

To install the repair anchor a 19-mm diameter clear hole was initially cored through the granite panel. A larger 30-mm diameter hole for countersinking the screw was cored approximately 6 mm into the face of the granite panel. The annular ring of granite between the clear and countersunk hole was chipped out. A 19-mm diameter hole was then drilled incrementally through the aluminum strongback extrusion. A toggle bolt was installed through the clear hole, with the toggle bearing against the inside face of the aluminum extrusion. A cap screw with thread locking compound was then threaded into the toggle bolt and tightened to a specified torque. See Fig. 8 for views of the installed anchor. Finally, a granite cover plug was placed over the countersunk cap screw and adhered with a limestone-colored silicone sealant (see Fig. 9). The selected silicone sealant is formulated specifically for sensitive porous stone substrates to reduce substrate staining.

Summary

A loose granite panel discovered during a routine facade inspection required a follow-up inspection to determine if it posed a structural concern. The initial inspection opening of this panel revealed that one of the two epoxied pins had failed and the second was close to failure. An exterior inspection, numerous structural analyses, an interior inspection, and proof load testing were all performed to determine the extent of epoxied pin connections that were improperly installed and inadequately designed. Once it was determined which panels had epoxied pins that could not adequately resist design wind loads, a repair was designed and implemented for those panels.

Journal of ASTM International, Vol. 4, No. 9
Paper ID JAI100842
Available online at www.astm.org

J. Eric Peterson[1] *and William H. Blodgett*[2]

Energy Savings Resulting from Building Envelope Upgrades in Mid-Rise Construction—A Case Study

ABSTRACT: The Bice House Dormitory on the campus "Grounds" of the University of Virginia was constructed in the mid-1970s as a dormitory for the nursing school. The building was an eight-story, concrete framed, residential facility with brick veneer cladding over steel stud framing systems. After more than 25 years of service, deficiencies in the brick façade, coupled with inadequate exterior wall framing, required removal and replacement of the entire exterior wall system from the interior drywall, up to and including the brick veneer. As part of the exterior wall replacement, window systems and insulation materials were upgraded. The building was also refitted with an air barrier system over the new exterior gypsum sheathing. Because existing individual packaged terminal air conditioning units were temporarily removed and reinstalled after renovation, the heating and cooling systems for the building remained unchanged, providing a unique opportunity to measure the true energy savings realized in a building strictly as a result of upgrading the building envelope. When the total heating and cooling costs after the first two full years of service were compared to the costs from years prior to the renovation (normalized for the temperature variations), a distinct reduction in energy costs were recognized from the improvements made during the renovation. The yearly energy consumption for the building was reduced by roughly 14 %, yielding a savings of roughly $12,800 per year based on current energy costs. Based on a payback period method of analysis, the improvements will have paid for themselves in roughly 14 years.

KEYWORDS: masonry, brick, veneer, window, insulation, air barrier, renovation, recladding, energy savings, building envelope

Introduction

The Bice House is an eight-story dormitory constructed in the mid-1970s on the campus "Grounds" of the University of Virginia (Fig. 1). It was originally used as a dormitory for the nursing school students and has been in continuous operation since construction. Currently, the dormitory provides housing for upper class students of all disciplines. The building was designed during a period in which brick veneer curtain wall systems were beginning to gain popularity among architects. The relatively thin and lightweight cross sections produced more interior space and wider hallways than equivalent concrete masonry back-up wall systems, resulting in greater interior area for similar or lower costs than in more traditional construction. Unfortunately, during the 1960s and 1970s, attention to details for accommodating differential movement was largely ignored by architects and contractors. Early generations of steel stud/brick veneer systems frequently experienced problems due to unaccommodated expansion in the masonry, excessive deflection of the veneer and back-up, and deterioration of the sheathing, anchors, and studs. Such was the case in the Bice House.

Background

During routine inspections of the building, University of Virginia Housing Division personnel noticed that minor vertical cracks near external corners in the brick veneer had become wider and they requested assistance from the University's Facilities Management organization. A closer examination of the veneer identified out-of-plane displacements and bulges at the shelf angle locations, in addition to the cracking at building corners (Figs. 2 and 3). In some cases, some of the out-of-plane displacements exceeded 1 in. and were potentially unstable. The original construction of the brick veneer did not include either vertical or

Manuscript received November 3, 2006; accepted for publication August 23, 2007; published online October 2007.
[1] Associate, Whitlock Dalrymple Poston & Associates, Inc., Manassas, Virginia.
[2] Project Manager, University of Virginia, Charlottesville, Virginia.

FIG. 1—*The Bice House (just prior to completion of renovation).*

horizontal expansion joints. The shelf angles at each floor were installed directly above the top of the masonry veneer from the floor below without adequate space to accommodate brick expansion and concrete frame shrinkage. Expansion of the brick veneer horizontally at perpendicular walls resulted in masonry stress at the building corners. The resulting restrained horizontal and vertical movement was resolved through out-of-plane displacements at the shelf angles and cracks at opposing planes of veneer; a schematic illustration of this condition observed at typical shelf angles is depicted in Figs. 4 and 5.

FIG. 2—*Out-of-plane displacement at window head.*

FIG. 3—*Cracking at building corner from out-of-plane displacement.*

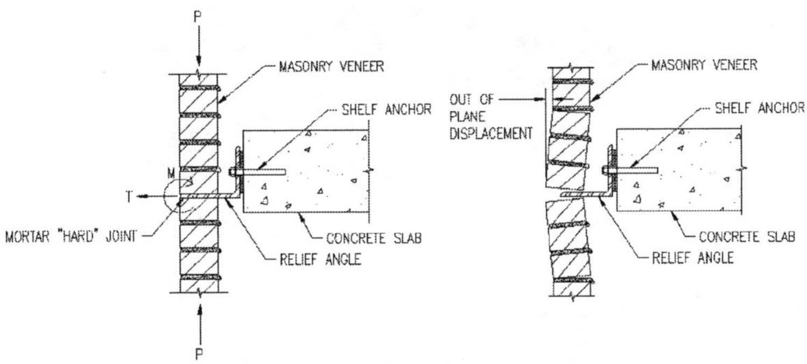

FIG. 4—*Schematic of restrained vertical expansion resulting in out-of-plane displacement at relief angles.*

An investigation of the brick veneer was performed to identify the cause of the distress and develop alternatives for the repairs. It was determined that in addition to the visible cracking and displacement, the entire masonry veneer was in generally poor condition with numerous deteriorated mortar joints, incorrectly installed and corroded brick ties, and corroding steel lintels and shelf angles (Figs. 6 and 7). The analysis of the structural capacity and serviceability performance of the original wall system determined that the steel stud walls did not have adequate strength or stiffness, nor were they adequately connected to the building frame. Substantial strengthening of the walls would have been required to meet the provisions of the modern building codes and industry recommendations.

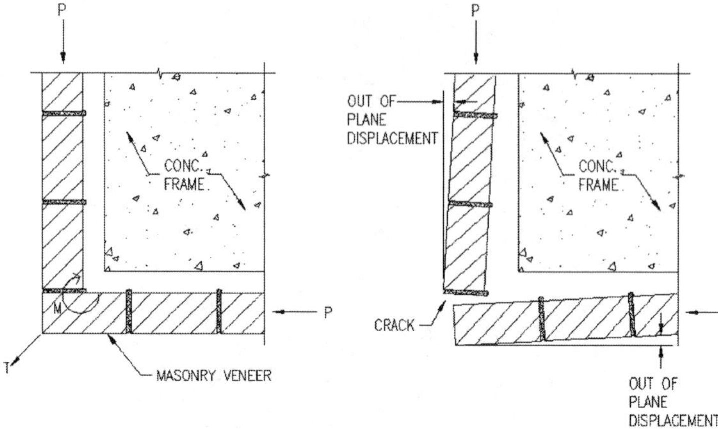

FIG. 5—*Schematic of restrained horizontal expansion resulting in out-of-plane displacement at building corners.*

FIG. 6—*Corrosion of loose lintel over PTAC unit louvers.*

FIG. 7—*Incorrectly installed and corroded corrugated brick tie.*

Numerous conditions were observed in the original construction that significantly impaired the performance of the building envelope. Aluminum window systems had neither insulated glass, nor thermally broken frames. Sliding sashes were often loose in the tracks with poorly sealed gaskets around their perimeter. The resulting air leakage through most of the window systems was quite evident. Additionally, there were numerous unfilled gaps in the gypsum sheathing and masonry veneer around wall penetrations such as vents, windows, and mechanical sleeves. In some cases, when drywall was removed from the interior, visible daylight could be observed from inside the building with the veneer still in place. Localized evidence of leakage and condensation-related damage to the exterior wall systems indicated that the building envelope was very loose and that the resulting air infiltration into the building contributed to the increased costs to heat and cool the building.

Three repair alternatives were evaluated to correct the deficiencies, including:

1. Localized repair of distressed portions of the veneer,
2. Replacement of the existing veneer with supplemental strengthening of the steel stud walls, or,
3. Replacement of the building's entire exterior wall system (masonry veneer and steel stud walls) taking the opportunity to perform a full building renovation as well.

It was decided that alternative #3 would provide the university not only with the opportunity to correct the deficiencies in the exterior masonry, but also to improve other elements of the facility, unrelated to the problems with the exterior walls, such as plumbing, interior finishes, electrical wiring, fire alarm systems, and other items as part of the entire construction project.

Project Description

The project included the removal of the entire exterior skin of the building, including the brick veneer, gypsum sheathing, steel stud walls, interior drywall, exterior windows and doors, and other related exterior wall components. The building's roof system had been recently replaced and remained in service throughout the project.

Prior to the design of the repair project, the university assembled a team of professionals including a forensic building consultant, an experienced architect from the University's Facilities Management Division, a professional construction cost estimator and a "local" experienced general contractor. The group was charged with evaluating methods of improving the thermal efficiency of the existing wall system with cost-effective alterations to typical design details and repair methods. The team identified several energy efficient improvements to the building envelope that could be implemented without prohibitively impacting the overall construction budget:

- An increase in the stud size from 3 5/8 in. to 6 in., an increase in the yield strength from 33 ksi to 50 ksi, an increase in the stud gage thickness from 20 ga. to 16 ga. and an increase in the stud spacing from 16 in. o.c. to 24 in. o.c. to reduce the number of conductive transmission points in the insulated wall and improve the overall strength and stiffness of the building exterior.
- A change in the wall insulation from foil-faced, $2\frac{1}{2}$ in., R-8 to unfaced, 6 in., R-19 fiberglass batt insulation.
- Installation of new heavy commercial (HC) grade aluminum window systems with insulated low E glass and a thermal break.
- Addition of a continuous air barrier system that was tied into penetrations and bridged across floor lines.
- Upgrades to the interior vapor barrier system to be continuous across framing and sealed at penetrations.

Electrical improvements, plumbing modifications, kitchen improvements, new fire alarm systems, and many new interior finishes were also added to the building, but the original room layout and the vast majority of the interior partition walls in the core of the building were preserved from the original design. Only the improvements related to the exterior wall system are believed to have had significant effect on the efficiency of the building envelope. Improvements to the exterior wall system were selected over and above that which were necessary to restore the structural capacity of the cladding. These improvements added an estimated \$175,000 to the total project cost of approximately \$5,000,000. The Fig. 8 sequence depicts the condition of the building before, during, and after construction.

One of the most unusual conditions of the renovation project involved the heating and air conditioning

FIG. 8—*Bice House before, during, and after renovations.*

systems for the building. Individual rooms within the building were equipped with packaged terminal air conditioning (PTAC) units below the windows. These units had been replaced four years prior to the renovation and the university's Housing Division requested that the units be removed, stored during the renovation, and then reinstalled in the building. Unlike many other whole building renovations, this case study is unique in the fact that by preserving and reinstalling the heating and air conditioning systems, and installing lighting and appliances with similar efficiencies to those removed, the effect of the building envelope improvements could be tied directly to the change in energy consumption when adjusted for climatic variations.

Energy Consumption Analysis

The University of Virginia's Energy and Utilities Division of the Facility Management Department maintained meter reading information for the building and provided summaries of the reading dates and the energy usage per monthly billing cycle. These data were evaluated for the seven-year period from January 1999 when the PTAC units were installed until July of 2006. Excluded from the analysis is the 15-month period of construction for the renovation from roughly May of 2003 until August of 2004, when the building's electrical service was generally disconnected except for construction power service. The resulting data sample captures four winter and five summer cycles before the renovation and the two winter and two summer cycles after the renovation.

Because the data from the meter readings were not consistently measured on the same date of each month, each monthly kilowatt-hour (kWh) reading was divided by the number of days between readings to get an average daily electric usage. It should also be noted that because the total energy usage was recorded, it includes power utilized for lighting, appliances, and other nonclimate control elements. However, as previously discussed, it is not expected that power consumption would be significantly altered by their replacement.

Also provided by the university's Energy and Utilities Division was the local climate data for Charlottesville, Virginia, in the form of heating and cooling degree-days over the same period. These data represent the difference between the mean daily exterior temperature and a baseline reference point, normally 65°F. The degree-day data were grouped by month and separated into heating degree-days (HDD) and cooling degree-days (CDD). Similar to the energy data, the monthly degree-day data for each seasonal period were divided by the number of days per period to develop average daily degree-day values.

The PTAC unit efficiencies differ between the heating and the cooling cycles and the occupancy of the buildings change significantly between the winter and summer semesters. Therefore the data were grouped into peak seasonal "heating and cooling periods" to minimize the effect of combined heating and cooling periods in the spring and fall months. November through March (five-month period) was taken as a heating season with little to no air conditioning usage. June through August (three-month period) was taken as the cooling season with little to no heating usage. By separating the data into solely heating or cooling periods of use, differences in the building envelope efficiency can be determined for both conditions. Table 1 depicts a summary of the data obtained by the university over the seven-year period.

The average daily kilowatt-hour usage was divided by the average degree-days to provide an average consumption rate per degree-day for each seasonal grouping. By comparing the consumption rates before and after the renovation, an estimate of energy consumption savings can be identified from the seasonal climate data. A more simplified calculation, which does not differentiate between the heating and cooling cycles of the PTAC units, divides the total heating and cooling degree-days for the year by the total energy consumption for the same period. This is used to compare the total energy savings before and after the construction. Table 2 depicts an overview of the findings from the energy consumption analysis.

Conclusions

Based on the data recorded from before and after the renovation of the Bice House, it is apparent that a considerable reduction in energy consumption has been recognized from the improvements made to the building envelope. Savings during the winter months, which are often more substantial than the summer months, demonstrated a dramatic reduction in the consumption per degree-day; generally about 30 %.

TABLE 1—*Seasonal energy data.*

Seasonal Period	kWh per Billing Cycle	Days in Billing Cycle	Average kWh per Day	Cooling Degree-days	Heating Degree-days	Days in Seasonal Period	Average Daily Degree-day
Nov 98-Mar 99	668880	153	4372		2892	151	19.15
July 99-Aug 99	421200	83	5075	1305		92	14.18
Nov 99-Mar 00	728640	155	4701		2928	152	19.26
July 00-Aug 00	295920	92	3217	757		92	8.23
Nov 00-Mar 01	759600	157	4838		3742	151	24.78
July 01-Aug 01	289440	95	3047	909		92	9.88
Nov 01-Mar 02	623520	155	4023		2799	151	18.53
July 02-Aug 02	347040	97	3578	1191		92	12.95
Nov 02-Mar 03	676080	154	4390		3880	151	25.69
Apr 03-Oct 04			Information Not Available During Renovation Project				
Nov 04-Mar 05	539280	151	3571		3553	151	23.53
July 05-Aug 05	302400	92	3287	1014		92	11.02
Nov 05-Mar 06	514080	160	3213		3389	151	22.44
July 06-Aug 06	180000	56	3214	597		61	9.79

TABLE 2—*Overview of the energy consumption analysis.*

Period	Average Daily Energy Consumption (kWh/Day)	Average Daily Degree-day Values (DD/Day)	Average Energy Consumption per Degree-day (kWh/DD)	Percent Savings
Cooling from '99–'02	3729	11.31	329.7	5 %
Cooling from '04–'06	3251	10.41	312.3	
Heating from '99–'03	4465	21.48	207.9	29 %
Heating from '04–'06	3392	22.99	147.5	
Total from '99–'03	4244	14.53	292.2	14 %
Total from '04–'06	3446	13.65	252.4	

Savings during the summer months were much less pronounced at about 5 %, but are still considered significant. Overall, the yearly savings on energy consumption averaged around 15 % for the facility. As expected, the effects of the improvements are most significant during the extreme warm and cold periods rather than during average temperatures.

The total cost of the building improvements was about $175,000.00 over the basic remediation work. Based on the estimated unit energy cost in 2006 of $0.06/kWh and an average 14.5 degree-days/day over the past ten years in Charlottesville, VA, it is estimated that more than $12,800.00 (2006 dollars) will be saved in an average year. Based on a "Payback Period Method" of analysis [1], the improvements will pay for themselves in roughly 14 years, well within the anticipated remaining service life of the building.

Often when faced with replacement of an "architectural" building system, a presumption is made that replacement in-kind is not only simpler, but the only practical way to accomplish the intended goals. Designers setting forth the additionaol effort to carefully analyze the physical constraints of the building, the anticipated benefits of the alternatives, the related additional construction costs, and the means and methods of the improvements, can often add significant value to the project. Although each repair/renovation effort is different, this project illustrates how carefully chosen energy-saving improvements can save significant utility costs. The savings recognized over the remaining service of the building can often far exceed the initial costs of the improvements during construction.

Reference

[1] White, J. A., Agee, M. H., and Case, K. E., *Principles of Engineering Economic Analysis*, John Wiley & Sons, Inc., 1984.

Journal of ASTM International, Vol. 5, No. 10
Paper ID JAI100881
Available online at www.astm.org

Warren R. French[1] *and Joshua S. Schroeder*[2]

Delayed Ettringite Formation (DEF) Within Pre-Cast Concrete Cladding Panels on a Mid-Rise Commercial Office Building: A Case Study

ABSTRACT: This paper will present the findings of multiple forensic investigations on a fourteen-story office building located in Houston, Texas. During our initial survey, widespread cracking was observed within the concrete cladding panels that warranted a more extensive investigation. Construction of the building was completed in 1985; 17 years prior to our initial investigation. Exterior cladding components are comprised of pre-cast concrete panels over each of the floor lines, with a continuous, structurally glazed, horizontal window system at each floor level. The cladding panels are nonload bearing, pre-cast concrete panels, installed at the time of the building's construction. The concrete panels are attached to the structure via steel plates that are welded and bolted to the building's frame. It is intended that this paper will present the results of observations made at the site over a period of several years, as well as the results of testing conducted to assist in the evaluation. A system was developed to locate and catalog the visible anomalies on the panels; including cracks, exposed steel, spalls, pop-outs, and efflorescence on the panels. Full-depth cores were removed from select panels to perform a petrographic analysis and scanning electron microscopy/energy dispersive x-ray microanalysis (SEM/EDXA). The results revealed the presence of delayed ettringite formation (DEF) and alkali-silica reactivity (ASR), with varying degrees of severity. Specific conclusions will be derived related to the conditions at this project and how these anomalies have affected the cladding system performance. Current repair and maintenance provisions will be discussed, in addition to the necessary recladding that will be required in the future. Recommendations will be developed regarding construction practices so that the cladding panel failure observed may be avoided in the future.

Introduction

In the summer of 2002, a preliminary assessment of the exterior cladding system of a mid-rise office building located in Houston, Texas was commissioned by the management company responsible for the building. The first phase of the assessment was conducted to determine the overall condition of the cladding system, which is comprised primarily of a glass and aluminum horizontal window system and pre-cast concrete spandrel panels. An additional purpose for the initial study was to determine the effectiveness of prior remedial work that had been conducted by a contractor six years previously. The findings of our initial cladding condition survey revealed a rather prevalent condition of horizontal and vertical cracking occurring within the pre-cast panels, with a few minor spalls. Recommendations of that initial study included, among other items, that additional investigation be conducted on this cladding system, including extraction of concrete cores, structural analysis, and laboratory testing.

During the second phase of the preliminary investigation, visual observations were made of the entire cladding system, including the exposed exterior surfaces of each pre-cast panel. In addition, full-depth cores were obtained and subjected to petrographic examination. Analysis of those cores revealed that the pre-cast concrete panels had experienced fairly significant and widespread delayed ettringite formation (DEF) within the concrete matrix of the panels, with a minor contribution to the deterioration from alkali-silica reactivity (ASR). The overall result of this ettringite formation was an increase in the internal stresses present within the concrete, causing internal micro-cracking, which lead to general weakening of the concrete and the observed cracking within the surfaces of the panels. A third phase of the initial investigation included eight drops using personal descent devices, which allowed inspection of eleven "stacks" of pre-cast panels (due to the saw-tooth configuration). As more of the panel surfaces were

Manuscript received October 25, 2006; accepted for publication October 16, 2008; published online December 2008.
[1] President, P.E., RRC, CCS, French Engineering, Inc., 4201 FM 1960 West, Suite 300, Houston, Texas, 77068.
[2] Regional Manager, E.I.T., French Engineering, Inc., 3 W. Garden St., Suite 205, Pensacola, Florida, 32502.

FIG. 1—*Typical building elevation.*

evaluated and compared to the core petrographic results, it became apparent that there did not seem to be a direct correlation in the condition of the concrete matrix as revealed by petrography and the results of observable anomalies, such as the number of cracks and spalls. Accordingly, we became convinced that a comprehensive visual assessment would be required of the entire building skin.

While previous remedial repairs had been attempted by the building manager in order to waterproof the panels at the cracks, those repairs were deemed to have been either insufficient or too late to prevent the formation and gradual progression of the ettringite crystals.

Building Description

The subject property is 14 story office building located in a northern suburban portion of Houston, Texas. While the footprint of the building is basically rectangular in shape, there are saw-tooth portions at the Southwest and Northeast corners of the building (Figs. 1 and 2). The occupied space of the building is approximately 2323 square meters (25 000 S.F.) of usable area per floor, with 33 120 square meters (356 500 S.F.) of gross rentable area, and the building is approximately 50 m (174 ft) tall at the top of the pre-cast parapet. The building is, and has remained, 100 % occupied, which presented unique problems for the required investigations and remedial repairs. The structure of the building consists of cast-in-place concrete columns and floor slabs.

Horizontal strip windows at this project consist of aluminum and glass, two-sided structurally glazed, custom curtain wall assemblies that have been incorporated into the cladding system at each floor level on all elevations of the building. The horizontal strip windows are composed of 1 in. thick insulating glass with decorative horizontal, snap-on covers at the head and the sill and low modulus silicone structural glazing at the vertical butt joints occurring between the individual glazing panes.

Original construction of this building occurred during 1986 and 1987 and the building was approximately 16 years old when the problems with the panels were first investigated by our company. A previous inspection by another consulting firm was conducted in 1994 (eight years prior to ours) for the purpose of a due diligence survey related to the purchase of the building. In addition, a limited exterior renovation was performed using swing stage inspections on the pre-cast concrete panels in 1996, consisting of routing and sealing certain visible cracks with elastomeric sealants. At the same time, a limited amount of exterior sealants, primarily related to the window perimeters, were removed and replaced on an as-needed basis. It is not known whether the cracks within the vertical surfaces of the panels were prevalent at that time; however, cracking at the head and sill edges of the panels was a major portion of the limited renovation performed at that time.

FIG. 2—*Building saw-tooth corner.*

Pre-cast Concrete Panels

There are a total of 420 pre-cast panels that comprise the exterior building envelope of this project. All of the pre-cast concrete panels are either 2.44 m (8 ft) tall or 3.05 m (10 ft) tall, depending on floor level. The North and South elevations both consist of 154 panels each, with lengths of the panels including 3.0 m (10 ft), 7.32 m (24 ft), 9.14 m (30 ft), and 10.0 m (33 ft), depending on location within the fenestration. Likewise, the East and West elevations both consist of 56 panels each, with lengths of the panels including 6.10 m (20 ft), and 14.0 m (46 ft). The most prevalent panel length was the 9.14 m panel (total of 150), occurring in both the 2.44 m and the 3.05 m heights. The longest panel types (14.0 m) occurred only in one stack at each of the East and West elevations (total of 28 panels).

The pre-cast panels at this project consisted of two different finishes on the exposed exterior surfaces, which were generally separated by recessed aesthetic reveals arranged both horizontally and vertically within the surface of the panels. The reveals had been cast into the panels with a trapezoidal profile during fabrication. Upper portions of the pre-cast concrete panels exhibited a sand-blasted, smooth surface finish, while the lower portions and ends exhibited an exposed aggregate surface finish. It is likely the exposed aggregate portions of the panels were originally fabricated using a retarding agent that allowed additional "working" of the finish subsequent to the forms being stripped. It should be noted that cracking within the panels at this project was more difficult to locate and follow at the surfaces finished with exposed aggregate, as compared to the sand-blasted surface. However, commonly, a crack observed within the smooth area could then be traced into the aggregate surfaced area for a certain distance. Although anomalies were observed at exposed aggregate locations, a vast majority of the anomalies recorded by our visual inspections were located within the sand-blasted, smooth-surfaced panel areas, and our report analysis did not attempt to distinguish between these two areas.

FIG. 3—*Concrete core with two layers.*

The cast-in-place floor slabs and perimeter beams typically included two recessed steel angle embeds for each panel that "matched" the locations of flat plate embeds cast into the backs of the panels. The connection between the pre-cast panels and the perimeter beam embed consisted of a steel angle bracket that was welded to both the panel and floor embeds, usually with steel shims utilized to adjust the panel for erection tolerances. These connections generally conformed to the eccentric bearing type designated by the Precast/Prestressed Concrete Institute (PCI) as 9EB6 [1]. Although this connection configuration allowed adjustment during construction, there was no provision within this assembly for thermal expansion and contraction. Accordingly, there would be stresses introduced within these pre-cast panels as a result of thermal expansion and contraction, depending on the distance between connection embeds and the temperature differential (Fig. 6). Lateral stability at the bottom of the panels was provided by a bolted diagonal angle brace (or tie-back) that PCI designates as BT1 [1]. Although no records were available from the original pre-cast fabricator, information obtained from sources familiar with original construction indicate that the fabrication and curing procedure for the panels provided to this project utilized specially constructed steam curing rooms in order to accelerate hydration and curing of the portland cement components of the panel mix. In addition, it was reported that this fabricator generally utilized relatively high curing temperatures for extended periods of time. It is now generally recognized that high temperature curing can retard or inhibit the natural formation of ettringite during the hydration process. If this natural process of ettringite formation is inhibited during the initial cure, it will generally occur after the primary hydration has taken place and will typically progress after the concrete panels are erected and in service. Several technical papers and articles have been written about the formation and detrimental effects of delayed ettringite formation (DEF). [2–5].

Observations at this project were performed using personal decent devices on each elevation, with four rappel drops made on each of the East and West elevations, and eight rappel drops made on each of the North and South elevations. Each panel was identified and numbered using a designation scheme with

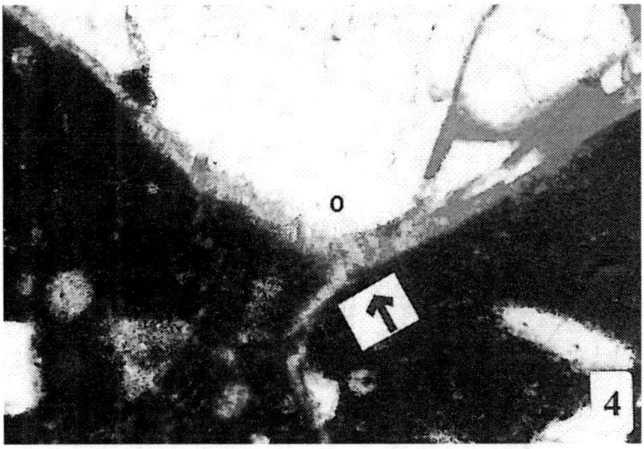

FIG. 4—*DEF by scanning electron microscope.*

letters and numbers to designate building elevation (N, E, S, and W), then floor level (1 through 14, and "R" for roof parapet panel), and finally the number of the drop designation reading left to right (1 through 4 on the East and West elevations and 1 through 8 on the North and South elevations). Accordingly, examples of panel designations would include N-11.1 (North elevation, eleventh floor, first panel from left), S-R.7 (South elevation, roof parapet, seventh panel from left), E-3.4 (East elevation, third floor, fourth panel from left), and W-14.2 (West elevation, fourteenth floor, second panel from left).

Descriptions of Delayed Ettringite Formation and Alkali-Silica Reactivity

Delayed Ettringite Formation

Ettringite is a chemical compound (calcium tri-sulfo-aluminate) found within concrete as it cures. When allowed to form naturally, it is part of the hydration and curing process that is beneficial for concrete performance. Delayed ettringite formation consists of a conspicuous late formation of ettringite in voids, cracks, and the contact zones between aggregate and the hardened concrete paste. DEF is generally seen as a form of internal sulfate attack, and the late reaction of high sulfate concentration in the pore liquid with calcium and aluminum containing phases of the cement paste results in the late, or delayed, formation of ettringite. It is generally recognized that DEF is associated with heat-treated pre-cast elements that were subjected to temperatures of 68°C (154 °F), or greater. In addition, the delayed formation of ettringite during concrete use is associated with exposure to open-air weathering with frequent wetting. In addition, DEF is mainly associated with high grade concretes of high strength and low porosity. The reaction causes expansion, resulting in cracks (gaps) around the aggregate and micro-cracking throughout the concrete paste (Figs. 3–5). Cracks may remain empty or later be partly or completely filled with ettringite. DEF is typically diagnosed by the presence of gaps encircling aggregate, wider gaps around larger aggregate, and high temperature curing history.

Alkali Silica Reactivity

Although only a minor portion of the deterioration mechanism at this project, alkali-silica reactivity (ASR) was found within a certain portion of the panels tested. ASR consists of the expansive deterioration of concrete due to a chemical reaction involving the components in aggregates and the cement paste. It is strongly influenced by the amount of sodium and potassium content within the cement, and the reaction generally causes higher pH (alkalinity) within the concrete. ASR is the result of siliceous aggregates reacting with the sodium and potassium hydroxides to form a silica gel, which is expansive. Moisture is required for the reaction and the problem is often associated with high alkali additives (i.e., certain fly ashes). When the silica gel expands, it creates internal stresses that eventually produce cracks in the

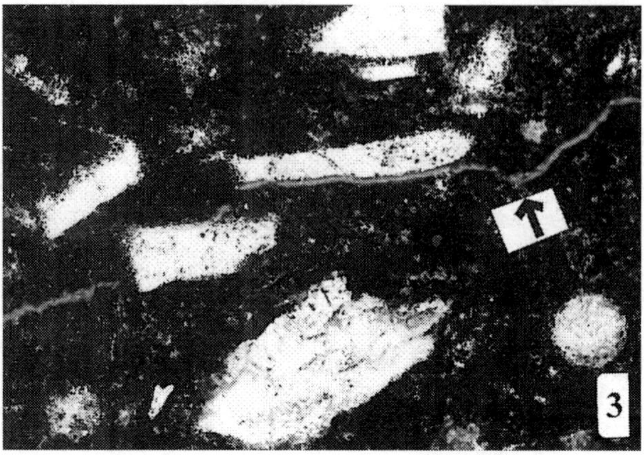

FIG. 5—*Micro-cracking within concrete.*

aggregate and the cement paste, which leads to physical distress. ASR is often distinguished by "alligator" cracking on the surface of the concrete, but may be inhibited at the exposed surfaces due to carbonation of the concrete [6].

Both DEF and ASR are best identified by petrographic examination in conformance with ASTM C856. Both DEF and ASR are progressive (particularly in the presence of moisture) and are essentially irreparable once the process has occurred. The progression of both can be inhibited by reducing exposure to moisture, alleviating outside sources of sulfates, and reducing introduction of outside sodium and potassium. In addition, some studies suggest that lithium compound applications after the fact have been effective in arresting the progression of ASR.

DEF can be avoided in the production of concrete panels by not allowing heat curing during fabrication. ASR can usually be avoided in the production of concrete panels by testing aggregates for reactivity prior to concrete mixing.

Investigations Conducted to Date

Initial Investigation

The initial investigation at this project was conducted in two phases with the first phase occurring during mid-July 2002. This first phase investigation consisted of three swing stage drops selected at random throughout the various elevations of the building. These drops were designated Drops 1, 2, and 3, respectively. The purpose of that survey was to ascertain the general condition of the building and to determine the effectiveness of minor renovation work performed several years earlier. The primary issues arising from the initial survey of the pre-cast concrete spandrel panels that exist at every elevation of this building involved the widespread and prevalent horizontal and vertical cracking observed within the concrete panels. However, other signs of deterioration were observed within the pre-cast panels, including spalls, alignment offsets, and exposed reinforcing steel and embedments.

At the time of the initial survey, and possibly due to the limited scope of the investigation, horizontal cracking within the pre-cast panels was observed to be more prevalent than the vertical cracking; however, we observed regular occurrences of both during our survey. The horizontal cracking was observed to range in size from approximately 0.15 mm to 0.40 mm (6 mils to 16 mils) in size and was typically found within 76 mm to 152 mm (3 in. to 6 in.) of the window sill (top of panel) on the front face of the pre-cast panels (reference Fig. 9). Our initial inspection revealed only one horizontal crack on the top floor of Drop 1 along the North elevation; however, similar cracking was observed on four of the panels of Drop 2 (31 %), and on nine of the panels of Drop 3 (69 %). The severe horizontal cracking at the panel edges

FIG. 6—*Cracking at embeds and bearing connections.*

involved 14 of the 36 panels (~39 %) observed during the three different drops of the initial survey of this building and caused significant concern regarding the condition of the unexamined portions of the building.

The vertical cracking observed during our initial site investigation was not as prevalent as the horizontal cracking but, in our opinion, is significant in the fact that the cracks observed were similar in size to the horizontal cracking, and spanned the entire height of the concrete panels in some instances. We initially observed vertical cracking on six of the panels (46 %) of Drop 1, three of the panels (23 %) of Drop 2 (one panel exhibited two vertical cracks), and no occurrences on Drop 3.

Another anomaly observed during our initial survey was the occurrence of what appeared to be "form" cracks at the panel edges adjacent to the sills and heads of the windows on the flat horizontal portions of the pre-cast panels. The cracking observed during our inspection appeared to be similar to the cracking that was repaired along the tops of the panels during the 1996 renovation. This cracking occurred at what appeared to be a "cold joint" between two pours of the panel concrete (gray and colored); however, we did not have any specific information on the manner in which these panels had been cast and the fabricator was no longer in business. We determined that these cracks were most likely caused by shrinkage stresses during curing. During the 1996 renovation, a majority of the cracks along the tops of the panels had been caulked; however, a number of untreated cracks still existed along the bottoms of the panels. In addition, it was noted that numerous locations along the tops of the panels were not adequately repaired during the 1996 renovation and still had the potential to allow water infiltration into the pre-cast panel, potentially causing corrosion of the underlying panel reinforcement.

During our initial survey, we also observed a number of locations throughout the building exterior where the pre-cast panels were exhibiting different types, and varying degrees, of spalling, the potential for spalling, or both (Figs. 7 and 8). Spalls present an aesthetically displeasing appearance on the exterior of the building, particularly one with sand-blasted and exposed aggregate pre-cast concrete. However, spalls of concrete from a high rise building also pose a safety issue if they occur above pedestrian traffic areas. It was known that spalls in concrete are generally the visual result of a release of internal pressure or stress within the concrete. This internal stress can originate from a number of different sources, including expansion and contraction of adjacent wall components, freeze-thaw cycles of entrapped moisture, corrosion expansion of underlying metal components, or expansion due to superheated moisture trapped within the pre-cast panel (often the result of a fire). Houston typically experiences winters that do not produce extreme freeze-thaw cycles and there had reportedly been no fire experienced by this building.

During the three mechanical swing stage drops made during our initial investigation, numerous horizontal and vertical cracks within the pre-cast concrete panels were observed. Since only a small portion of the building was observed during the initial investigation, and due to the fact that we were unable to determine by visual examination the cause of the prevalent cracking observed, we recommended that further investigation of the widespread cracking in the pre-cast panels be performed to determine the actual

FIG. 7—*Corner spall at panel joint.*

extent of damage to the panels, and the necessary remediation requirements. The investigation would include review of panel shop drawings, structural calculations, petrographic examination, and a review of panel thermal expansion and contraction analysis.

Subsequent to the initial survey and prior to the first comprehensive survey, our company conducted an inspection of three additional drops using swing stage equipment. During this second phase of the initial investigation, conducted late June of 2003, we conducted additional visual inspections and made arrangements for extraction of six concrete cores through the entire depth of the pre-cast panels. These cores were subjected to comprehensive petrographic examination and energy dispersive X-ray analysis. The full discussion of these test results is presented below. In summary, it was found that the pre-cast concrete panels had experienced fairly significant and widespread delayed ettringite formation (DEF) within the concrete matrix of the panels, with a minor contribution to the deterioration from alkali-silica reactivity (ASR). The initial test results were confirmed by extracting additional cores and having additional petrographic examination conducted by a second laboratory.

FIG. 8—*Spall at outside corner of panel.*

FIG. 9—*Typical horizontal crack at top of panel.*

Analysis of Cores

Of the six full-depth cores removed from the panels in July 2003, five were tested using petrographic analysis, and four were tested using scanning electron microscopy/energy dispersive X-ray microanalysis. The samples were prepared by cutting two sections from each core that was tested. One section was cut from the exterior face of the core, and continued down into the core for a depth of approximately 3 in. The other section was removed near the interior face of the panels to determine how deep the chemical attack had progressed. A blue dye was vacuum impregnated during preparation to make micro-cracks and voids in the paste more visible, and to determine the water-cement ratio (W/C) with a reasonably accurate degree of certainty.

Scanning electron microscopy is a form of high resolution microscopy that utilizes electrons rather than visible light. The electrons create pulses that are scanned as images that create a high depth of focus and very high magnifications ($\times 10^3$). In addition to the electrons, X rays are produced and detected. The resulting X rays can be differentiated between elements that are present within the cores. This type of X ray analysis is called energy dispersive X ray analysis and can be quite useful in determining the constituent elements of a compound.

The resulting analysis determined that the visible cracks were wide, but relatively free of chemical attack. However, the walls of the cracks were carbonated to a certain depth. In addition, the paste was found to be relatively dense, resulting in a low W/C ratio (below 0.45) and a high cement factor. The concrete is also air entrained, with air content ranging from 6 % to 8 % on average.

The air voids present within these samples were small (<0.2 mm) to medium (0.2 mm to 0.5 mm), and were often found to be partially or completely filled with ettringite crystals. No micro-cracks were observed that were associated with the ettringite filled voids. The coarse aggregate is siliceous in nature (mostly chert), and the fine aggregate is a combination of quartz sand and highly angular minerals.

The DEF observed within the concrete cores from this building had mainly formed at the paste-to-aggregate interface. While DEF was also observed within the entrained air voids, at the time of the initial petrographic examinations, it was not deemed to be harmful to the overall performance of the panels. However, the formation of DEF at the paste-to-aggregate interface was determined to be detrimental to the panels by acting to force apart the concrete from the inside of the panel. Research suggests that DEF initially forms at the paste-to-aggregate interface, forms micro-cracks, and then spreads throughout the concrete.

Calcium hydroxide was also found at the interface between the paste and aggregate. The calcium hydroxide is reportedly consumed during the chemical reaction in which the DEF is formed. The consumption of the calcium hydroxide results in a highly porous rim along the micro-crack. While the paste was observed to be very dense throughout the samples analyzed, a porous zone was found to be present on both sides of micro-cracks containing deleterious DEF.

ASR was also found to exist in approximately half of the cores tested, primarily in the exterior portions of the core. It is the opinion of the author, and of others who have reviewed the analysis, that the ASR present within the concrete cores of this building is a secondary reaction that has resulted from the formation of the DEF. The interior portions of the cores were found to be in acceptable and serviceable conditions, with the exception of one core.

First Comprehensive Survey

The first comprehensive survey of all pre-cast panels at this project occurred in December of 2003 and consisted of 24 drops conducted by engineering personnel using personal descent devices (i.e., no swing stages) for which they had been trained. This method of inspection was chosen for the ease and rapidity of setting up and moving the ropes from location to location on the building, as well as lower overall costs and the decreased potential for causing damage to the building.

Based on our visual observations, at least one crack existed at approximately 94 % of all of the panels that were observed at this project. Approximately 97 % of the concrete panels at this project are affected by at least one anomaly, including cracks, exposed reinforcing steel, concrete pop-outs, or spalled areas. Approximately 4 % of the concrete panels at this project contained cracks that were 0.3 mm or greater in width, which is the ACI tolerable crack width for concrete exposed to weather [7].

The second phase of the initial investigation had previously identified certain types of accelerated concrete deterioration occurring within the pre-cast concrete cladding panels of this project. The most prevalent type of concrete deterioration being experienced is the delayed ettringite formation (DEF), which has been confirmed by two different rounds of panel coring, petrographic examination, and scanning electron microscopy analysis performed by two different laboratories. The presence of DEF within the various cladding panels varied widely, from virtually no effect in some panels to severely affected and advanced in other panels. The second type of concrete deterioration identified within the cladding panels is alkali-silica reactivity (ASR), which has been characterized as being not as prevalent, nor as advanced, as the DEF problem. It has been noted that both of these accelerated deterioration mechanisms are progressive over the life of the building, and both are exacerbated by water exposure and moisture infiltration. The comprehensive visual survey implemented during the first comprehensive study was intended to document the condition of the panels at the time and to discern patterns of deterioration, if any, that may be visible to inspection. In addition, it was intended that this survey would serve as a "baseline" for future surveys.

Due to the geographic location of Houston, buildings in this area receive a majority of exposure to the sun on the South and West elevations. The predominant wind direction in this location is from the South and Southeast directions, and a majority of the rainfall in this area is experienced on the same elevations. As a result, it is has been our experience that cladding deterioration at projects in this region are typically more prevalent on these elevations. Based on the information obtained during the initial site investigations at this project, cracks and anomalous conditions also appeared to be more prevalent on the South and West elevations. Approximately 67 % of all cracks observed on the outer faces of the panels, and 63 % of all cracks observed, including along the top and bottom edges of the panels, occurred on the South and West elevations at this project. These elevations represent approximately 50 % of the total cladding system area and essentially two-thirds of the panels exhibiting serious anomalies. Excessive cracking (more than three cracks per panel) and panels with cracks that were 0.3 mm or greater in width, were also more prevalent on the South and West elevations. Furthermore, cracks appeared to be slightly more prevalent at the lower halves of each building elevation rather than the upper halves, which may be attributable to the cumulative effect of cascading water during and subsequent to weather events.

We observed that a large number of cracks within the pre-cast panels at this project correlated with underlying steel reinforcing elements that were either corroded, installed with insufficient concrete coverage, or a combination of these factors. Other cracks observed at this project may also be attributable to internal forces present within the panels due to their anchor assemblies, which have been installed with a "fixed-fixed" type assembly (i.e., no provision for lateral expansion and contraction).

Based on a comparison between the number of cracks previously observed within various panels on certain elevations of the building and an analysis of the results of laboratory testing conducted by two different laboratories, we concluded that there does not appear to be a strong or direct correlation between the severity of cracking visible at that time and the severity or extent of DEF and ASR identified by sample extraction. In other words, our analysis indicated that the amount of cracking observed at any one panel may not be an indicator of the presence of DEF or other deterioration mechanisms. Accordingly, it appears that a continuing program of sample extraction and laboratory testing is the only practical method to further assess the extent and severity of the deterioration being experienced. Based on those conclusions, it was our recommendation to the building manager that future, on-going, periodic visual inspections at this building will be required in order to (a) provide advanced notification and identification of potentially dangerous situations, (b) provide further monitoring of the potentially progressive deterioration of the original pre-cast concrete cladding panels, and (c) to compare future inspection results to the baseline sketches provided within the first comprehensive survey.

Our investigation indicated that at least one crack occurred on 94 % of all panels observed at this project. As previously noted, and based on our experience with similar projects in this geographic region, deterioration is commonly more prevalent at the South and West locations due to greater exposure to weather elements. Non-crack anomalies, such as concrete pop-outs, spalls, and efflorescence, were also more prevalent at these elevations.

As noted in the past partial surveys of this building, horizontal cracks appeared to be more prevalent than vertical cracks during our previous site investigations. However, the first comprehensive survey revealed that the total number of vertical cracks were slightly more frequent than horizontal cracks. It should be noted that many of the horizontal cracks were uninterrupted and spanned greater distances than vertical cracks (although vertical cracks could have continued unnoticed through exposed aggregate locations). Horizontal cracks could therefore be more prevalent on a lineal foot basis. Practically speaking, it is our opinion that vertical and horizontal cracks, on average, are relatively consistent in frequency of occurrence and have had proportional detrimental affect on the concrete panels at this project. Both horizontal cracks and vertical cracks were also commonly observed on the same panels.

During this investigation, a large number of locations at this project were observed where cracks corresponded directly with corroded reinforcing steel embedded within the panels. While some locations exhibited reinforcing steel that was exposed due to inadequate embedment into the concrete, a number of concrete pop-outs were observed where the reinforcing steel appeared to have been adequately embedded. It is our opinion that reinforcing steel observed at this project has deteriorated due to chronic moisture infiltration at a number of locations, including deteriorated sealant joints, cracks, and absorption through the surface of the porous panels themselves.

First Follow-up Partial Survey

One of the primary purposes of the first comprehensive survey was to establish a baseline for the condition of each of the panels. Subsequent to that survey, the building manager desired to assess the progression of the DEF and cracking based on a partial investigation in order to avoid the costs of another comprehensive survey. The first follow-up survey of this building included rappel inspections performed on select panels of the building in August of 2004 in order to track the progression and formation of cracks and anomalies on the surface of those selected panels. This interim inspection consisted of eight drops on the four elevations of the building, resulting in a total of eleven sets of panels being inspected. The saw-toothed configuration of the building and the short panel length allowed observation of two sets of panels at some of these drops. Three sets of panels were inspected on the North elevation, two sets on the East elevation, four sets on the South elevation, and two sets on the West elevation. A total of 154 individual panels (approximately 36 % of the total) were inspected during this phase of the investigation.

Of the 154 panels inspected in August 2004, 71 % had at least one crack on the front face of the panel, compared to 67 % in December 2003. This represented a 6 % change in just an eight month period. In addition, 18 % of the panels inspected in August 2004 had more than three cracks, in comparison to only 8 % of the panels inspected in December 2003 (a 125 % increase). Six percent of the panels investigated in December 2003 had cracks greater than 0.3 mm, compared to 12 % of the panels (double) inspected in August 2004. Analysis of the August 2004 data led the authors to conclude that the condition of the panels was worsening, and should be monitored regularly in order to continuously evaluate the condition of the panels. We also recommended that the building owner begin to develop plans for major repair efforts or else re-cladding of the building.

Second Comprehensive Survey

The second comprehensive panel investigation was conducted during January of 2005, approximately one year after the first comprehensive survey and about 17 months after the initial survey. The second comprehensive survey revealed a dramatic increase in the number of panels that exhibited either new cracks or else previous cracks that have increased in length. This anomaly was consistently observed on all four elevations. On the West elevation of the building, 82 % of the panels exhibit new cracks, cracks that have increased in length, or both. Although graphical representation of the panel data from this survey as compared to the first comprehensive survey was developed, in the interest of space, we will wait until later in this paper to present these comparisons. In addition, 61 % of the panels on both the South and East elevations exhibited new or increased cracks, and 18 % of the panels on the North elevation exhibited new or increased cracks. Panels on each of the four elevations also exhibited increases in the number of top and bottom edge cracks, as well as non-crack anomalies. At the time of this survey, at least one crack existed within 95 % of all of the panels at this project (a 1 % increase). Approximately 98 % of the concrete panels at this project were affected by at least one anomaly, including cracks, exposed reinforcing steel, concrete pop-outs, or spalled areas. Approximately 14 % of the concrete panels at this project contained cracks that are 0.3 mm or greater in width.

During this survey, approximately 70 % of all cracks observed on the outer faces of the panels occurred on the South and West elevations at this project (an increase of three percentage points). Many of the original anomalies observed with respect to exposed reinforcing steel, corrosion of the embedded steel, and non-crack anomalies continued to be observed.

In light of the rapid increases in cracks and other anomalies on the panels on all four elevations of the building, the authors recommended that previous suggestions related to possible sealing of the panels with elastomeric coating be re-evaluated. One option that had been recommended previously consisted of sealing the cracks on all panels, and only removing and replacing the panels that were severely deteriorated and a liability to the public. However, based upon increases in both total cracks and non-crack anomalies between the first and second comprehensive surveys, we recommended that consideration be given to accelerating the planned removal and replacement of the panels. In addition, we recommended that previously discussed safety measures regarding the entry canopies be implemented, and that barricading around the building at the ground level be installed to avoid pedestrian traffic close to the building.

Based upon the increase in both the amount and severity of cracks and anomalies present within the panels, it was apparent that a continuing program of sample extraction and laboratory testing would be required to further assess the extent and severity of the deterioration being experienced until comprehensive removal and replacement of the pre-cast panels could be accomplished. In addition, it is our opinion that future comprehensive visual inspections at this building should be increased to once each six months in order to (a) provide advanced notification and identification of potentially dangerous situations, (b) provide further monitoring of the progressive deterioration of the original pre-cast concrete cladding panels, and (c) compare future inspection results to the sketches and analysis provided in this report.

Although cracks in the exposed aggregate surfaces were difficult to detect and follow, we observed several locations where cracks that were previously detected in the exposed aggregate surface were longer than previously recorded (Figs. 10 and 11). In addition, we observed several locations where cracks were present in the exposed aggregate surfaces where no cracks were observed during our previous inspections. A majority of the new and extended cracks observed in the exposed aggregate surfaces were observed in panels located on the West elevation; however, this condition was present on all four elevations of the building to some extent.

During our second comprehensive investigation, we observed a condition on the exposed aggregate corners of several panels on the West elevation that are of special concern. On four different panels located on the West elevation, we observed that a corner of each of the panels occurring within the exposed aggregate areas had deteriorated significantly, that and they were in danger of falling off. On many of these panels, there were hairline and 0.1 mm cracks at the upper-outside corner of the exposed aggregate. Since none of this cracking was observed during our previous inspections, it appeared that these cracks had formed and propagated in a short period of time.

During this inspection, we noted that at least one crack occurred on 95 % of all panels observed at this project, compared to 94 % of the panels observed in December 2003. In addition to the 1 % overall increase in the number of panels that exhibit at least one crack, we also observed that a majority of the

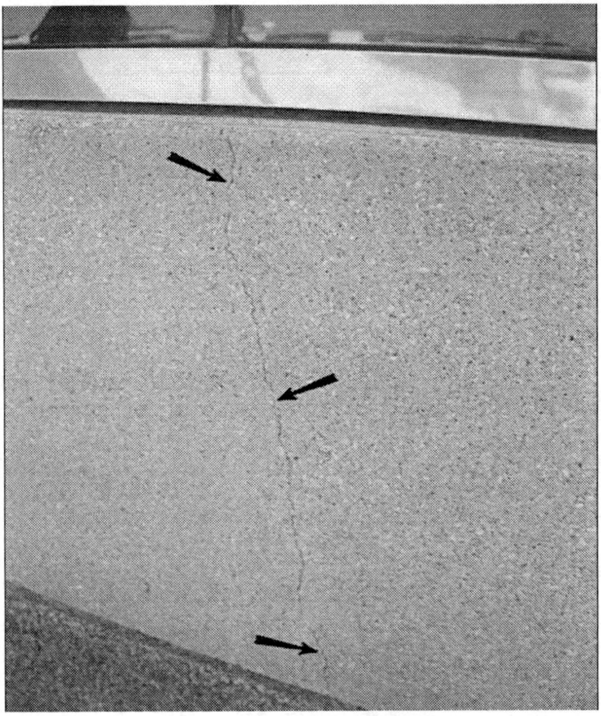

FIG. 10—*Typical vertical cracking.*

FIG. 11—*Close-up view of vertical cracking.*

panels on all four elevations of the building had either new cracks, or cracks that had increased in length. On the West elevation, 82 % of the panels had either new cracks or previous cracks that had increased in length from our previous site investigations. On the South and East elevation, 61 % of the panels had new cracks or increased crack lengths, and on the North elevation 18 % of the panels had new or increased cracks lengths. New top edge and bottom edge cracks, as well as new non-crack anomalies were present on panels on each of the four elevations of the building.

Our observations from the second comprehensive survey indicated that 59 % of the panels on the West elevation exhibited excessive cracks, compared to 14 % of the same panels in December 2003. This constitutes a difference of 45 % in approximately one year's time. In addition, the three other elevations also showed increases in the number of panels with excessive cracks.

Through analysis of the data obtained during our most recent building survey, we observed that cracks and anomalies, on average, were slightly more prevalent at lower floor levels opposed to upper floor levels. Overall, the cracks on the upper half of the building had an average of 2.07 cracks per panel, compared to an average of 3.12 cracks per panel on the lower floors. In December 2003, there was an average of 1.07 cracks per panel on the upper floors of the building, and 1.90 cracks per panel on the lower floors. This condition is another indication that the pre-cast panels have continued to deteriorate subsequent to our original investigations.

We observed an increase in the total number of horizontal cracks from 278 in December 2003 to 587 in January 2005. This is an increase of approximately 111 % in 13 months. The total number of vertical cracks increased from 288 in December 2003 to 396 in January 2005 (38 %). Based on these observations, it would appear the horizontal cracks have been recently increasing at a much greater rate than the vertical cracks. The increase in the number of horizontal cracks observed is likely the result of a combination of the internal crystallization of the concrete and corrosion of the embedded steel reinforcement observed during the previous surveys pertaining to this project.

The second comprehensive investigation also revealed a total of 1090 observable cracks on the front faces of the panels, compared to 623 cracks in December 2003. This constitutes a 75 % increase in the number of cracks observed on the building in a 13 month period. Dramatic increases were also observed in the number of panels that exhibited either new cracks, or previous cracks that increased in length. On the West elevation, 82 % of the panels exhibited new cracks, cracks that increased in length or both. The South and East elevation panels exhibited increases of 61 %, and the North elevation panels had an increase of 18 % in the number of panels that had new or increased cracks.

Third Comprehensive Survey

Based on the results of the third comprehensive survey, cracking within the concrete panels at this project continues to be more widespread and consistent on all elevations. At least one crack was found to exist at approximately 97 % of all of the panels that were observed at this project. Approximately 98 % of the concrete panels at this project are affected by at least one anomaly, including cracks, exposed reinforcing steel, concrete pop-outs, or spalled areas. Approximately 20 % of the concrete panels at this project contain cracks that are 0.3 mm or greater in width. Furthermore, the number of cracks that were observed to be 0.3 mm or wider increased significantly on the South, West, and North elevations of the building, increasing 94 % over the previous survey.

Our third comprehensive survey revealed an increase in the number of panels exhibiting either new cracks or previous cracks that have increased in length on all four elevations. On the East elevation, the average number of cracks per panel increased by 22 % from the January 2005 investigation (just seven months prior). Three of the four elevations exhibited a double digit percentage increase in the average number of cracks per panel, and the average cracks per panel for the entire building increased 12 % in this seven month period. Significant increases were also observed in the number of panels that had three or more cracks per panel, with the East elevation exhibiting a 40 % increase in the past seven month period. The non-crack anomalies were also observed to be more prevalent on most of the building.

Although cracks in the exposed aggregate surfaces were difficult to detect and follow, we observed numerous locations where cracks that were previously detected in the exposed aggregate surfaces were longer than previously recorded. A total of 34 panels were observed with this condition, with 10 panels on the West elevation, 13 panels on the South elevation, one panel on the North elevation, and 10 panels on the East elevation. In addition, we observed numerous locations where cracks are present in the exposed

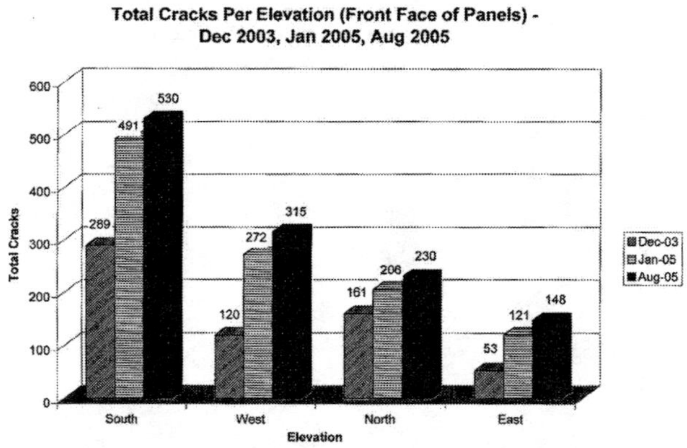

FIG. 12—*Total number of cracks by elevation.*

aggregate surfaces where no cracks were observed during our previous inspections. In total, 29 individual panels exhibited this condition. A majority of the new and extended cracks observed in the exposed aggregate surfaces were observed in panels on the West elevation; however, this condition was present on all four elevations of the building to some extent.

During the second comprehensive survey, we observed a condition on the exposed aggregate corners of several panels on the West elevation that were of special concern. On four different panels located on the West elevation, we observed that a corner of each of the panels occurring within the exposed aggregate areas had deteriorated significantly, and that they were in danger of falling off. The third comprehensive survey revealed four additional panels on the West and East elevations that exhibited this condition. On many of these panels, there were hairline and 0.1 mm cracks at the upper-outside corner of the exposed aggregate. Since the cracking in the corners of the four new panels was not observed in our previous inspections, it appeared that these cracks have formed and propagated in the seven months occurring between the second and third surveys.

Analysis of data from the third comprehensive survey indicates that at least one crack occurred on 97 % of all panels observed at this project, compared to 95 % of the panels observed in January 2005, and 94 % in December 2003. In addition to the 2 % overall increase in the number of panels that exhibit at least one crack, we observed that a majority of the panels on all four elevations of the building had either new cracks, or cracks that had increased in length. On the West elevation of the building, 45 % of the panels had cracks that increased in length, and 32 % of the panels on the South elevation had cracks that

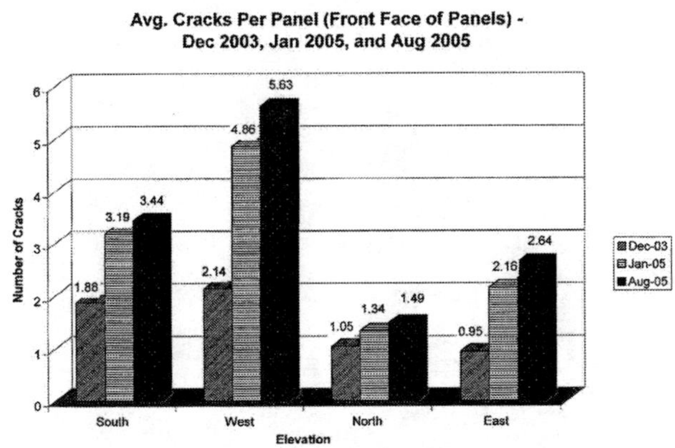

FIG. 13—*Average number of cracks per panel.*

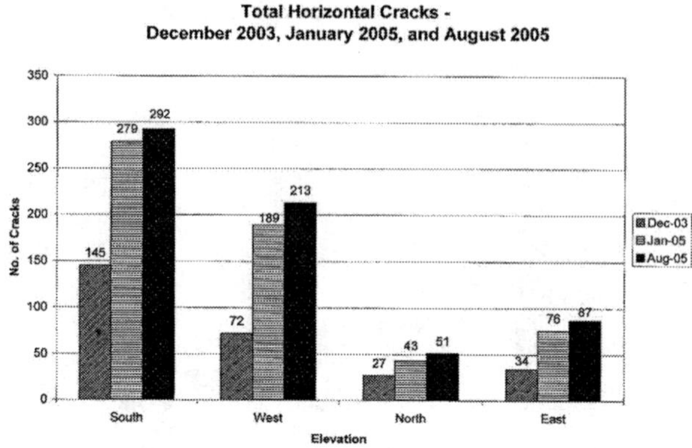

FIG. 14—*Total number of horizontal cracks.*

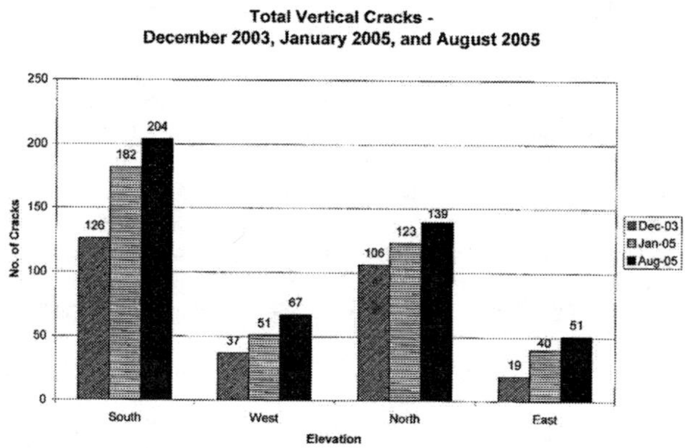

FIG. 15—*Total number of vertical cracks.*

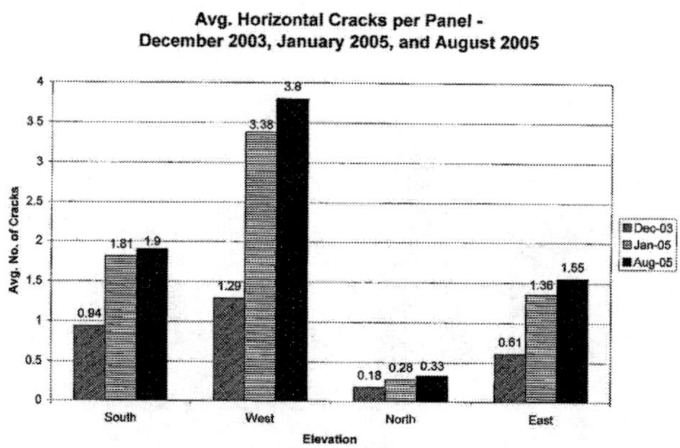

FIG. 16—*Average number of horizontal cracks per panel.*

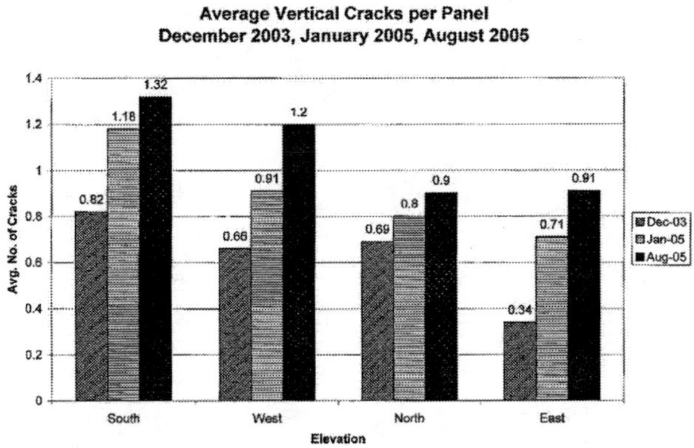

FIG. 17—*Average number of vertical cracks per panel.*

increased in length. Overall, approximately 24 % of the panels on the building had at least one crack that increased in length during the seven month period between the January 2005 survey and the current investigation.

As illustrated in Figures 12–17, most cracks, on average, occur on the South and West elevations. As previously noted, based on our experience with similar projects in this geographic region, deterioration is commonly more prevalent at these locations due to greater exposure to weather elements. We observed that non-crack anomalies, such as concrete pop-outs, spalls, and efflorescence, were also more prevalent at these locations. Overall, approximately 70 % of cracks observed on the front vertical face of panels were observed on the South and West elevations, which represent only 50 % of the total wall surface area. Furthermore, excessive cracking (more than three cracks per panel) and panels with cracks that were 0.3 mm or greater in width, were also more prevalent on the South and West elevations (Figs. 18–21).

During the third comprehensive survey, we observed that 66 % of the panels on the West elevation exhibited excessive cracks, compared to 59 % of the same panels in December 2003, and 14 % in December 2003. This constitutes a difference of 12 % in approximately seven month's time. In addition, the three other elevations also showed increases in the number of panels with extended cracks (Fig. 22). It should be noted that 25 % of the panels on the East elevation now have excessive cracking, up 40 % from the January 2005 survey.

During the third comprehensive survey, we also observed additional examples of non-crack anomalies on panels, such as spalls, pop-outs, and exposed reinforcing steel, occurring on all four elevations of the building. These anomalies were observed on the finished surfaces, the exposed aggregate areas, and the top

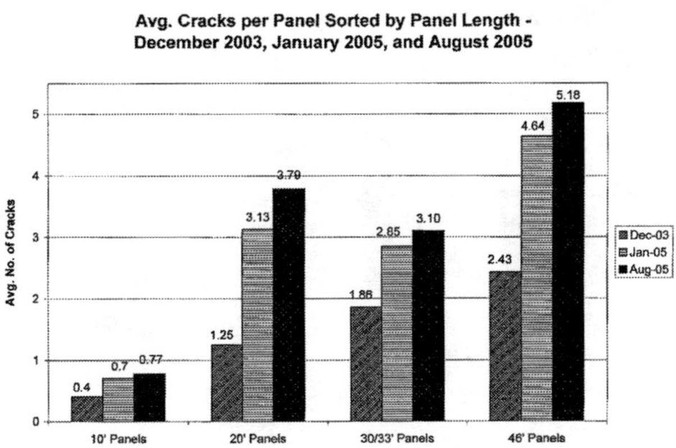

FIG. 18—*Average number of cracks per panel sorted by panel length.*

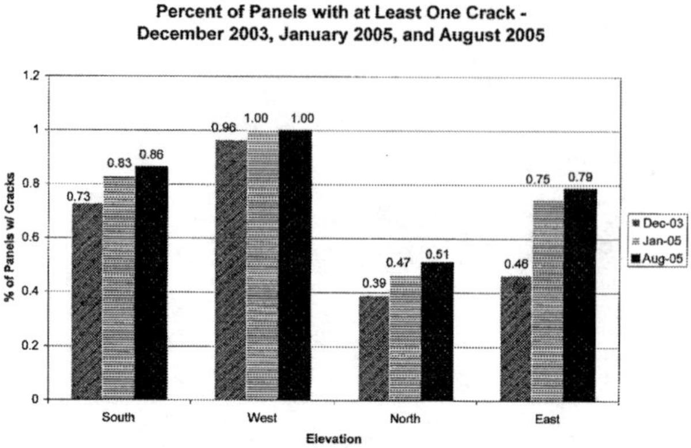

FIG. 19—*Percent of panels with at least one crack.*

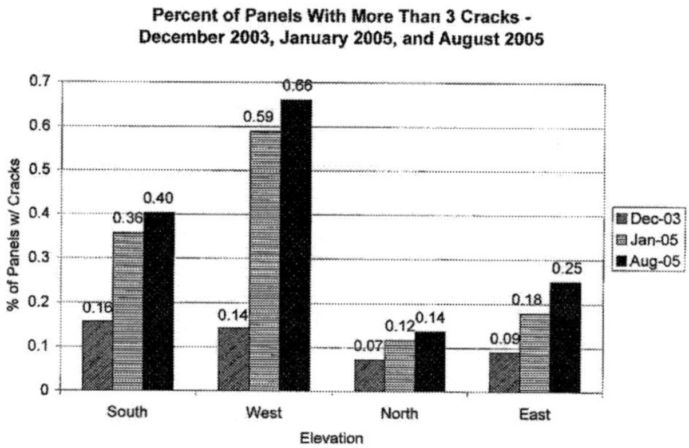

FIG. 20—*Percent of panels with more than three cracks.*

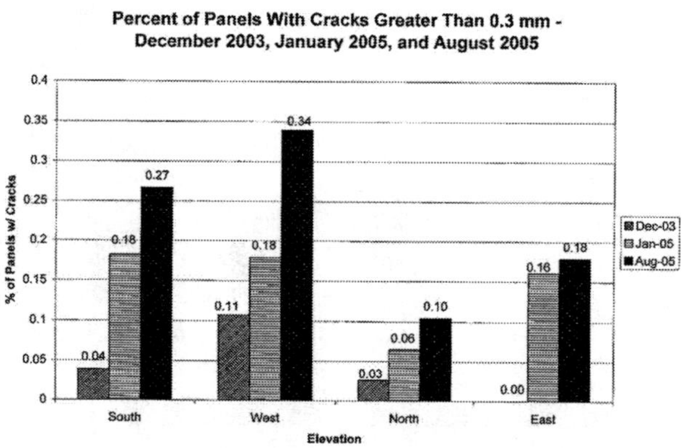

FIG. 21—*Percent of panels with cracks greater than 0.3 mm.*

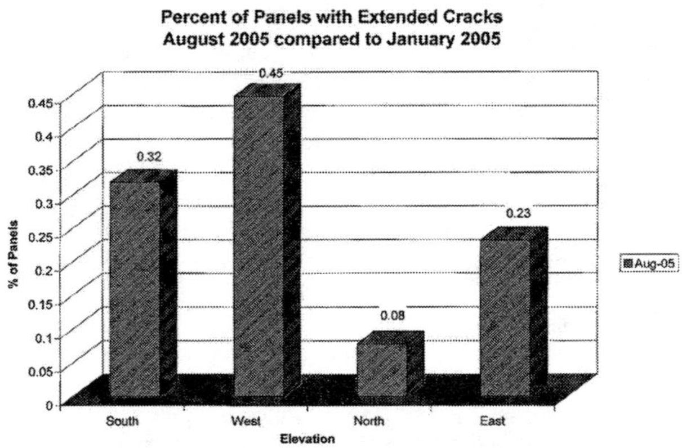

FIG. 22—*Percent of panels with extended crack.*

and bottom edges of the panels. New non-crack anomalies were noted in several locations on each of the four elevations of the building. These locations are evidence that the concrete panels have continued to manifest progressive deterioration subsequent to our previous inspections.

Based on an analysis of the data obtained during our third comprehensive survey, we observed that cracks and anomalies, on average, were slightly more prevalent at lower floor levels as opposed to upper floor levels. Overall, the cracks on the upper half of the building exhibited an average of 2.33 cracks per panel, compared to an average of 3.50 cracks per panel on the lower floors. In January 2005, there was an average of 2.07 cracks per panel on the upper floors of the building, and 3.12 cracks per panel on the lower floors. Our panel survey from December 2003 revealed 1.07 cracks per panel on the upper floors, and 1.90 cracks per panel on the lower floors. This average crack progression is another indication that the pre-cast panels have continued to manifest progressive deterioration subsequent to our original investigation.

As stated in the previous reports related to this project, the frequency of cracks on the outer face of the panels consistently increased in correspondence to the length of the panel, with more than six-and-a-half times the frequency for the longer (14 m) panels as compared to the shorter (3 m) panels (Fig. 18). As expected, the percentage of panels that had at least one crack, as well as the average number of cracks per panel, increased from the shorter panels to longer panels. In our opinion, the increased frequency of cracks observed as the panel length increases is due to the significant thermal expansion and contraction experienced by the longer panels. This condition may also be due to the special support and constraint conditions imposed on these panels.

The average number of cracks per panel continued to increase from January 2005 to August 2005 in all four of the panel length categories analyzed (3, 6.1, 9.14/10, and 14 m). However, the increase was not as significant as the increase observed from December 2003 to January 2005. For the 3 m (10 ft) long panels, the average number of cracks increased 11 %, while the average number of cracks for the 6.1 m (20 ft) long panels increased 21 %. Increases of 9 % and 12 % were observed in the average number of cracks for the 9.14/10 m (30/33 ft) long panels and 14 m (46 ft) long panels, respectively. In addition, substantial increases in the number of panels with more than three cracks were observed in the 6.1, 9.14/10, and 14 m (20, 30/33, and 46 ft) long panels, considering only seven months had passed since the previous survey.

In our original evaluation of the exterior cladding system at this building, we stated that, in the limited number of areas observed, horizontal cracks were more prevalent in the concrete panels than vertical cracks. However, our December 2003 comprehensive cladding survey revealed that vertical cracks were slightly more prevalent than horizontal cracks. Our January 2005 comprehensive investigation revealed at that time horizontal cracks were again more prevalent than vertical cracks, with the horizontal cracks increasing by 111 %, while the vertical cracks had increased by only 38 % from the December 2003 survey. Data from the August 2005 investigation revealed that the horizontal cracks continued to be more prevalent, and that the horizontal cracks had increased by 9 %, while the vertical cracks had increased by 17 % in the seven months intervening between inspections.

While many of the cracks observed on the subject panels were within acceptable tolerances for crack widths, numerous cracks are present that exceed the acceptable limits. In addition, we observed a substantial increase in the width of cracks as compared to those observed during our previous on-site investigations. In December 2003, 4 % of all of the panels present had cracks that were wider than 0.3 mm, and in January 2005, 14 % of the panels had cracks wider than 0.3 mm. Based upon data obtained during the third comprehensive survey, 20 % of the panels now have cracks that are wider than 0.3 mm, comprising one-fifth of the cladding system at this building. In our opinion, and based upon the data obtained in the most recent investigation, the cracks that were present on the panels in our previous investigations have increased in size in many locations, and will most likely continue to do so unless appropriate remedial actions are undertaken.

In our December 2003 comprehensive site investigation, we observed that there were slightly more vertical cracks than horizontal cracks on the outer faces of the panels. However, our January 2005 and August 2005 site investigations have revealed that there are now a much greater number of horizontal cracks than vertical cracks on the same panels. In our opinion, the vertical cracks that were previously observed on the panels are most likely the result of internal stresses in the concrete caused by the anchor conditions, and most likely started propagating not long after the panels were installed. After the panels cracked from the anchor conditions, the water that intruded within the panels caused the reinforcing steel to expand. The expanding steel then caused the concrete to crack further, which then allows more water inside the panels.

Our analysis of the data obtained in our investigations indicates significant correlations between the number of cracks and anomalies present within the panels and several factors, such as building elevations and locations on the building. In general, not only did the South and West elevations of the building exhibit more cracks and anomalies than the North and East elevations, but they also averaged more cracks and anomalies per panel. The number of cracks and anomalies on the North and East elevations appeared to have been growing at a significant rate as well. Additionally, we observed that the cracks are more prevalent on the lower half of the building, as well as on the in-set panels as compared to the out-set panels.

In lieu of complete removal and replacement of the pre-cast concrete cladding panels, we had previously recommended that all cracks within the panels should be sealed with an elastomeric coating and comprehensive remediation will be required where severely deteriorated panels are hazardous to the public. Based on the increase in anomalies and apparent progression of the deterioration within the cladding panels during the subsequent surveys, we have recommended that this approach be re-evaluated in light of the increases documented. In our opinion, consideration should be given to accelerating the planned removal of these defective panels, as well as the design, fabrication, and erection of a comprehensive cladding renovation at this project. Furthermore, it was our opinion that consideration should be given to barricading the immediate area around the building at ground level to avoid pedestrian traffic.

We have recommended that comprehensive panel inspections be continued at a rate of one inspection every six months using personal descent devices until comprehensive removal and replacement of the pre-cast panels may be accomplished. Comparisons may be made with the baseline sketches previously provided in order to continue to monitor and mitigate worsening conditions. It has also been our recommendation that it would be prudent to apply an elastomeric coating on all of the panels in order to alleviate future water penetration. However, the building managers have been notified that application of a pigmented elastomeric coating, while necessary to mitigate moisture effects, will make it more difficult to perform comprehensive visual inspections of the concrete panel surfaces to the degree that has been achieved on past inspections.

Based on the results of the third comprehensive survey of the concrete panels at this project, it is clear that DEF was still a major contributing factor to crack formation at this project, although there does not appear to be a distinct and definite correlation between the number of cracks and the presence of DEF within the concrete panels, as determined by laboratory testing. Accordingly, it is our opinion that periodic inspections should be conducted and that they include additional core sample testing and associated laboratory analysis, in order to further evaluate and monitor DEF progression within these panels. This course of action would require a combination of both rappel inspections and periodic swing stage investigations on all four elevations of the building in order to accomplish the additional core sampling and laboratory analysis.

Summary and Conclusions

This study was conducted over a three year period and included visual inspections, documentation by photographing and sketching of the panel appearance, core extraction, petrographic analysis, testing via scanning electron microscopy, infrared imaging, and a thorough analysis of the condition results and trends. We conducted partial visual inspections in July of 2002, June of 2003, and August of 2004. We also conducted comprehensive surveys of every panel in December of 2003, January of 2005, and August of 2005. The results of each of these inspections was thoroughly documented and compared to the same panel observations of the previous reports, with significant changes being noted.

We noted that a majority of the cracks within the concrete panels were within tolerable width limits, which may not warrant immediate comprehensive remediation, but rather an intermediate level of remediation in order to mitigate the widespread cracking. However, it is our opinion that comprehensive renovation (complete removal and replacement of the panels) should be performed at isolated locations that are severely deteriorated, including areas where crack widths are 0.3 mm or greater, and at locations where portions of the panels may be in danger of falling.

Previous condition surveys have noted an apparent correlation between the location of the cracks and the location of steel reinforcing members within the pre-cast concrete panels. During our comprehensive visual surveys, we observed this condition to be consistent on all elevations. Rusted reinforcing steel was observed to be a major contributing factor to crack formation at this project, and will continue to worsen at an accelerated rate as additional unsealed cracks create additional point sources for moisture infiltration into the panels. We observed the formation of new potential spalls at a number of locations during our third comprehensive survey, some of which are a potential danger to pedestrian traffic below.

Based on our analysis, a correlation exists between the building elevations and the severity and pervasiveness of anomalies observed. Cracks, pop-outs, and exposed reinforcing steel are more prevalent on the South and West elevations, and panel conditions at these locations may continue to deteriorate at a faster rate. Similarly, deterioration may be more advanced at lower floor levels due to cascading water and drying inefficiencies and at panels that are in-set from the plane of the wall, possibly due the additional support restraints applied at these locations.

Comprehensive renovation will eventually be necessary at all panel locations, which will require complete removal and replacement of the panels. During the interim period for this project, we have recommended comprehensive panel inspections on at least an annual basis using personal descent devices or swing stages. Comparisons should continue to be made to the baseline sketches originally developed, in order to monitor and mitigate worsening conditions. In our opinion, comprehensive periodic inspections are an invaluable, proactive approach to effectively monitor the evolving concrete panel conditions at this project, and may also prolong the longevity and serviceability of the panels by providing immediate recognition of subsequent cracking and deterioration.

In addition to the cracks and anomalies presented within this report, it is our opinion that the continued deterioration of the panels observed in our surveys is generally the result of the delayed ettringite formation (DEF) and alkali-silica reactivity (ASR) formation identified in our previous investigations. As water continues to intrude into the interior components of the concrete panels, the DEF and ASR crystal formation will inevitably progress. As the crystals continue to propagate, the internal stresses within the concrete will increase until more cracks are formed in the concrete. When additional cracks form, more water can enter into the panels, causing further crystal formation, in conjunction with rebar corrosion expansion. Once the critical crack limit is reached within the panels, they will begin to lose their inherent structural properties, and portions of the panels may begin to fall off of the building. Complete removal and replacement, or at the very least remedial measures, will need to be implemented on the subject building in order to mitigate the cracking and deterioration observed.

In our opinion, the corrosion and expansion of the embedded reinforcing steel occurring within the concrete panels continues to be a problem on the subject building. Through several on-site investigations, we have observed what appears to be a direct correlation between the location of horizontal cracks on the outer face of the finished surface and the underlying reinforcing steel. On numerous panels, there are horizontal cracks located approximately 3 in. to 4 in. from the top horizontal edge of the panel. Although we were not provided shop drawings for the pre-cast panels, our experience indicates that this is a typical depth at which reinforcing steel would be located within these concrete panels. When the concrete panels crack and allow water to migrate within the panels, the moisture will cause the steel to corrode. This corrosion causes the steel to expand, greatly increasing the internal stresses within the concrete, which is already weakened by DEF processes. The result of the increase in internal stresses is often manifested through more cracks in the surface of the concrete, which can allow more moisture to migrate within the panels, causing a cyclical deterioration process.

Based on this study, consultants involved in the investigation and repair of existing pre-cast concrete cladding systems must be aware of the potential harmful effects of delayed ettringite formation (DEF) and alkali-silica reactivity (ASR) in order to properly ascertain the presence of these conditions and to test for their effects. A systematic program of observations and recording of data must be established in order to properly document the panel conditions and so that comparisons may be made with observations from subsequent inspections.

Based on the results of this survey, design professionals involved in the specification of new pre-cast concrete cladding systems must be aware of the potential harmful effects of delayed ettringite formation and alkali-silica reactivity in order to properly specify safeguards against these anomalies within the cladding systems of future projects. Various entities have proposed and are using established pre-construction tests for aggregate to determine the potential for reactivity and these measures can be helpful in avoiding future problems with ASR. However, it is our understanding that steam curing of pre-cast panels is still used by some fabricators, and specification restrictions pertaining to temperature may be necessary in order to minimize the detrimental aspects of this practice.

Acknowledgments

The authors wish to express appreciation to members of the engineering staff of French Engineering, Inc. at the time of these surveys and analyses, including Jeff Garrison, Brian McDade, Joseph Ferrell, and Andrew Creamer. In Memoriam: Petrographer and professional colleague Dr. Shondeep L. Sarkar, Ph.D., P.E. of Sarkar & Associates, Inc. of Houston, Texas.

References

[1] Freedman, S., *Architectural Precast Concrete*, Precast/Prestressed Concrete Institute, Chicago, IL, 1989, pp. 208–209.

[2] Stark, J., and Bollman, K., "Delayed Ettringite Formation in Concrete," Bauhaus University, Weimar, Germany.

[3] Lerch, W., "Effect of SO_3 Content of Cement on Durability of Concrete," PCA Research and Development, 1945, 0285, 9 pp.

[4] Odler, I., and Gasser, M., "Mechanism of Sulfate Expansion in Hydrated Portland Cement," *J. Am. Ceram. Soc.*, Vol. 71, No. 11, 1984, pp. 1015–1020.

[5] Mehta, P. K., "Mechanism of Sulfate Attack on Portland Cement—Another Look," *Cem. Concr. Res.*, Vol. 13, 1983, pp. 401–406.

[6] Stokowski, S. J. Jr., and Sarson, J., "Alkali-Silica Reactivity in Concrete," Stone Products Consultants, http://members.aol.com/crushstone/flyasr.htm, September, 2006.

[7] ACI standard 224R-01, "Control of Cracking in Concrete Structures," American Concrete Institute, Farmington Hills, MI, 2001, Table 4.1, p. 224R-19.

Journal of ASTM International, Vol. 5, No. 3
Paper ID JAI100899
Available online at www.astm.org

Robert Bateman[1]

Sill Pan Flashing for Block-Frame Windows in Recessed Concrete Openings—Case Studies

ABSTRACT: Integrating sill pan window flashing can be challenging using block-frame windows or installing flashings within a recessed opening in a concrete wall. Two case studies summarize the details of design and retrofit installation for sill pan flashing of replacement block-frame window into the perimeters of existing recessed concrete wall openings. Provisions for sill pan drainage were included with technical solutions for two case studies. The design and detailing were also conceived to account for blending with the context of the existing architecture and minimizing the aesthetic concerns for the appearance of any exposed perimeter opening flashings. These case studies demonstrate detailing methodologies for wood and metal block-frame window frames and concrete wall openings applicable for new and existing exterior wall systems.

KEYWORDS: barrier-wall, block-frame window, concrete wall, design, detailing, flashing, recessed opening, reticulated foam weep, sealant joint, sill pan flashing, window

Introduction

Block-frame (non-fin) windows can pose problematic conditions to effectively seal against water intrusion when installed into concrete wall openings. These common installations typically rely only on perimeter sealant joints between frame and opening, which can fail in time. Block-frame windows are often installed flush to the exterior wall face to eliminate retained moisture against the sealant joints where recessed openings may have flat sills. These types of window frames and the concrete openings have been intended to perform as a "barrier-wall" drainage system with no supplemental provision, such as, a pan flashing, to drain the window frame or the recessed opening perimeter at the sill condition.

In our experience, building designers do not consider incorporating sill pans with block-frame windows because the window frame joinery seals and perimeter opening sealant are usually assumed reliable and durable. Sill pans under a window sill and visibly extending to the building exterior are also thought by designers as aesthetically distracting at the least and ugly at the most.

This paper reviews two examples of block-frame windows in concrete openings using sill pan flashings to supplement the performance of the perimeter opening sealant and to present an acceptable visual appearance where exposed to the exterior.

- The first case study offered the challenge of modernizing a 30-year-old college library with rusted and leaking steel window frames. Replacement aluminum window frames with energy efficient glazing were installed into the existing concrete wall openings. The recessed installation included a special metal sill corner flashing that extended minimally to the wall exterior and integrated with the perimeter frame-to-wall opening sealant joint.
- The second case study presented the opportunity of rehabilitating a decommissioned military base building for educational use. Replacement wood windows were retrofitted into existing recessed concrete openings, which had been previously closed. New sill pan flashing consisting of sheet metal copper flashing with specially fabricated "fins" was designed to attach to the wood frame and fit within a retrofit receiver groove saw cut into the concrete opening perimeter. The critical sill corner detail integrated the fin flashing with a full sill pan flashing while remaining concealed from exterior view.

Manuscript received November 9, 2006; accepted for publication February 14, 2008; published online March 2008.
[1] Staff Consultant, Simpson Gumpertz & Heger, Inc. (SGH), San Francisco, California.

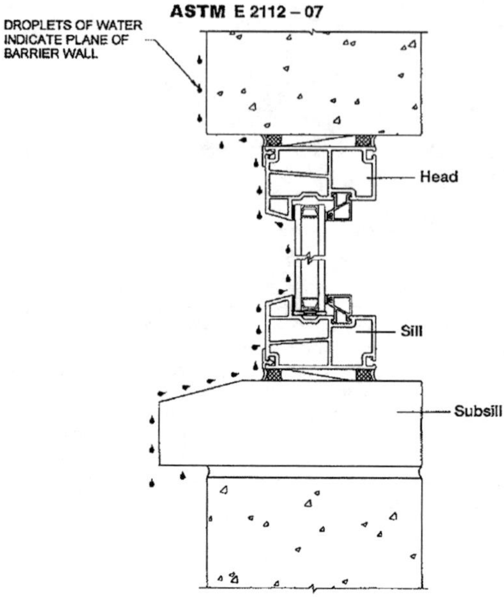

FIG. 1—*Concrete wall with block frame window in recessed opening [6].*

Performance Criteria

The performance criteria for barrier-wall systems are to prevent moisture intrusion at the surface of the exterior wall. Block-frame windows are the predominant window frame type used for barrier-wall systems. A perimeter sealant joint is typically used between frame and opening (see Fig 1). Most block frame window systems in barrier walls deflect water directly to the exterior. Hollow metal or vinyl frames may accommodate some incidental moisture penetration at the window frame, but manage water internally to drain outside through weep holes [1].

The following case studies present two different window types that manage potential water penetration with additional sill pan flashing that directs water directly to the exterior of a barrier-wall system [2].The sill pan flashing included side dams and a rear turned-up leg [3,4]. The perimeter opening sealant joint along the sill included a reticulated foam weep [2].

Case Study # 1—Alameda College Library, Alameda, California

The existing steel frame windows were replaced with aluminum frames in the existing openings during deferred maintenance repairs completed in 1998.

The original 30-year-old steel windows had exterior steel glazing stops that had failed over time, allowing water to enter around the stops into the hollow metal window frames. The retained water in the steel frames leaked to the building interior and eventually rusted through the metal section. Recurring water intrusion into the frame joinery allowed moisture to the floor slab and deteriorated interior finishes of the two-story concrete building (see Fig. 2).

Design Criteria

Aluminum frames matching the overall pattern of the original steel frames were selected as replacement frames with contemporary energy-efficient glazing (see Fig. 3).

To prevent future window frame leakage in the recessed openings, consideration for a sill pan flashing was included in the repair program. Minimizing changes in the exterior building appearance was also part of the repair criteria. The extension of the sill pan end dams was minimized and painted to match the

FIG. 2—*Rusted steel frames of original construction with exterior glazing stops.*

bronze anodized window frames (see Fig. 4). Since the building was situated in a temperate heating and cooling climate, the continuation of the metal sill pan flashing through the wall opening did not introduce a concern for thermal bridging [3,5].

FIG. 3—*West side of College of Alameda Library with replacement windows.*

FIG. 4—*Window sill-jamb corner with end dam extension of sill pan.*

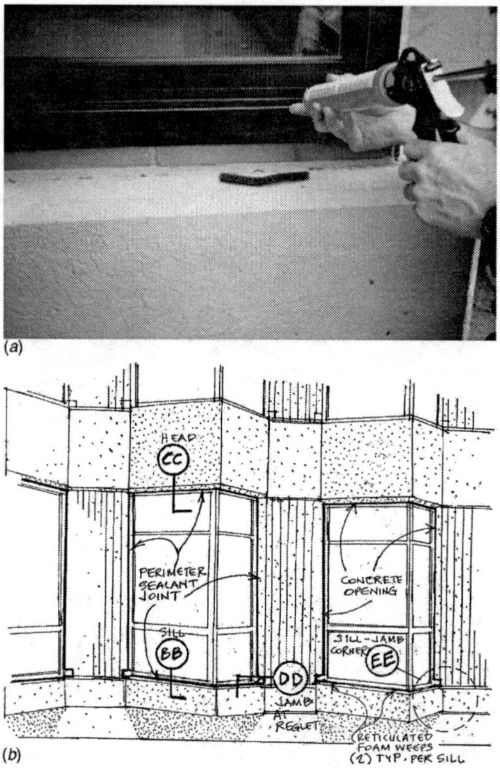

FIG. 5—*(a) Reticulated foam weeps at sill sealant joint. (b) Detail AA. Schematic window elevation.*

Details

New sill pans were added at the base of the windows. The repair included an elastomeric coating covering the existing, recessed concrete sills. Window perimeter details developed for head, jamb, sill, and corner included the components for flashing. See Figs. 5–10 details AA, BB, CC, DD, EE, and FF below for the new window design. The sill design incorporated an interior steel angle frame support to allow for vertical

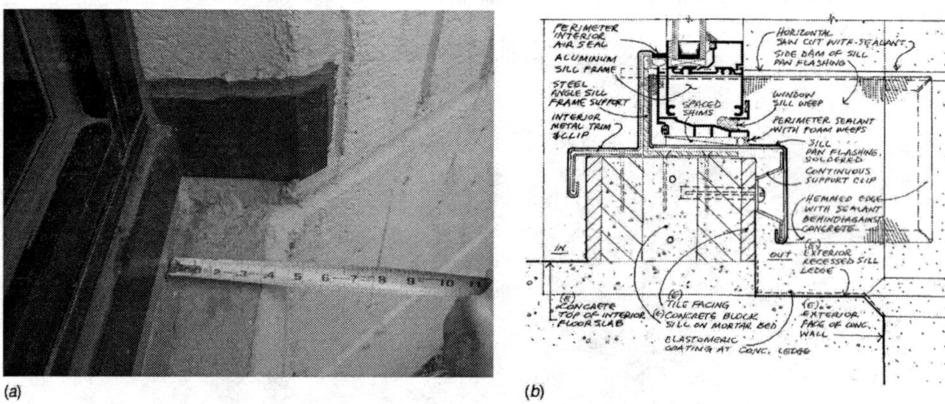

FIG. 6—*(a) Window sill. (b) Detail BB. Sill flashing.*

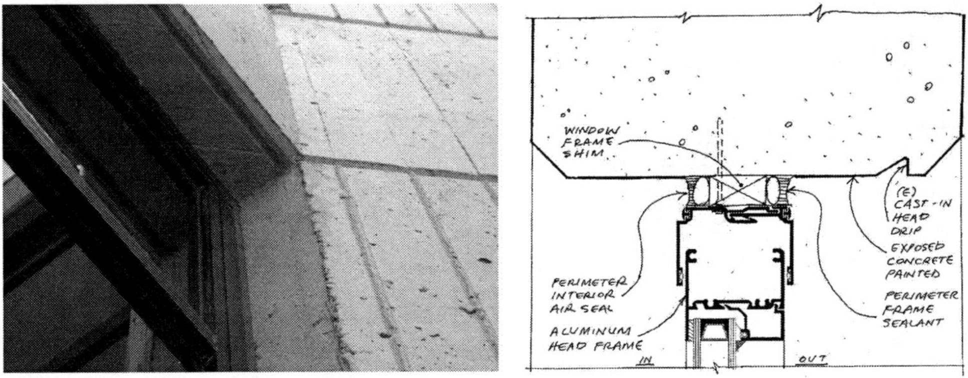

(a) (b)

FIG. 7—*(a) Window head.(b) Detail CC. Window head.*

fastening of the sill pan and thereby eliminating any fasteners in the horizontal plane [1,2]. The elimination of fasteners in the horizontal plane also ensured that the underlying sill pan flashing was not punctured and therefore provided a continuous unperforated back-up pan to the window subsill. The perimeter sealant at the sill was interrupted with weeps to allow the pan to drain (see Fig. 5).

Case Study #2—Lander Hall, Touro University, Vallejo (Mare Island), California

The former military building was completely rehabilitated in 1999 for educational use. The existing window openings were reopened after obsolete infill was removed and replacement wood-frame windows

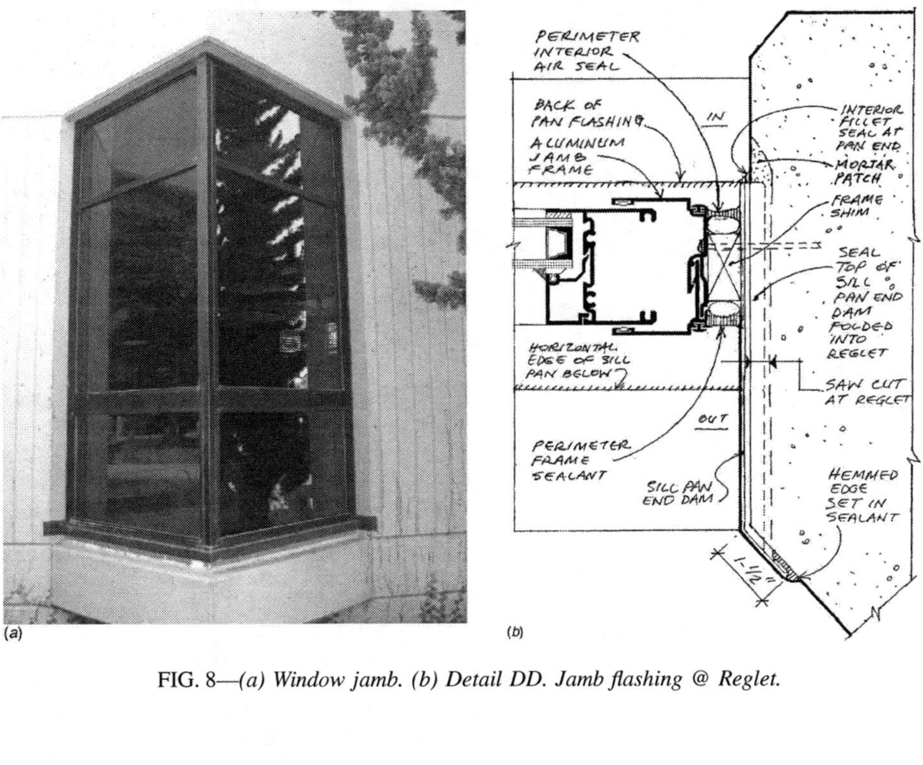

(a) (b)

FIG. 8—*(a) Window jamb. (b) Detail DD. Jamb flashing @ Reglet.*

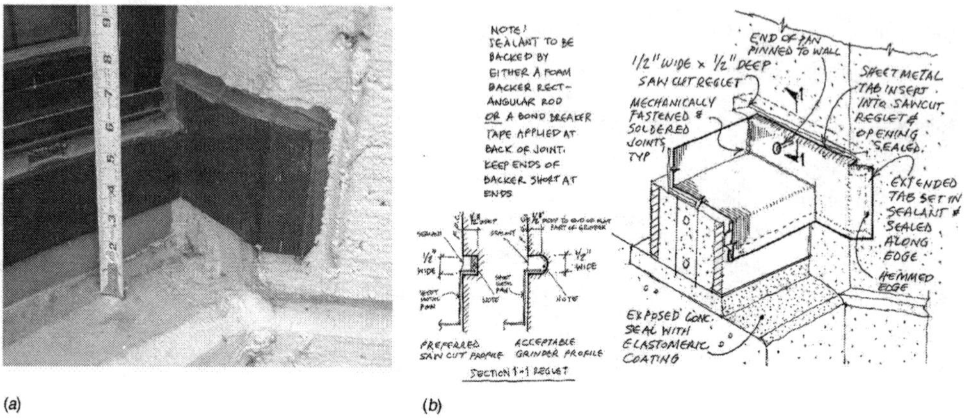

FIG. 9—*(a) Sill-jamb corner. (b) Detail EE. Sill-jamb corner.*

were installed. Figures 11 and 12 show the present 2006 exterior with the new windows in the existing openings.

The 1940s (World War II) era military building was utilized up until the Mare Island Naval Shipyard was decommissioned. The original three-story concrete building had most of the original wood windows removed and the openings were infilled to enclose the building walls for visual security during its last military use.

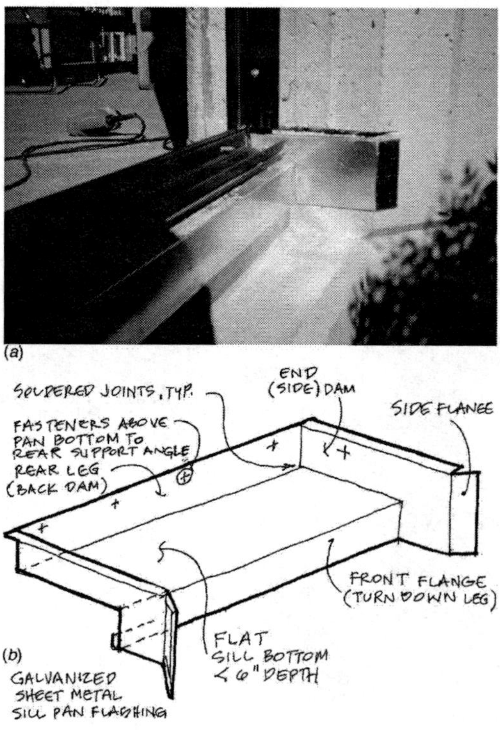

FIG. 10—*(a) Sill pan flashing. (b) Detail FF. Sill pan flashing.*

FIG. 11—*Southwest corner of Touro University with retrofit window openings.*

Design Criteria

New block-frame, wood-frame windows were installed into the existing, recessed concrete openings. The replacement windows included in the repair design were to appear appropriate to the style of the building's original windows. Painted replacement wood windows were selected for the rehabilitation of the building to new classroom use by Touro University (see Fig. 12). Additional perimeter window flashing behind the perimeter sealant joint was included with repairs.

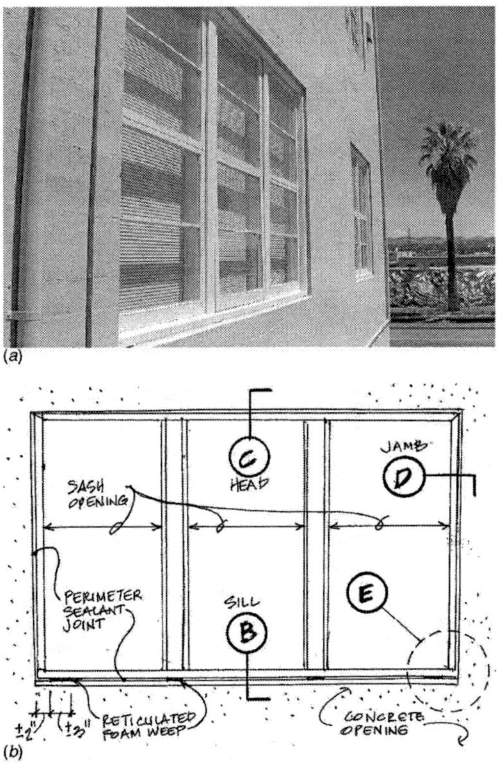

FIG. 12—*(a) Replacement wood-frame windows with perimeter sealant joints. (b) Detail A. Schematic window elevation.*

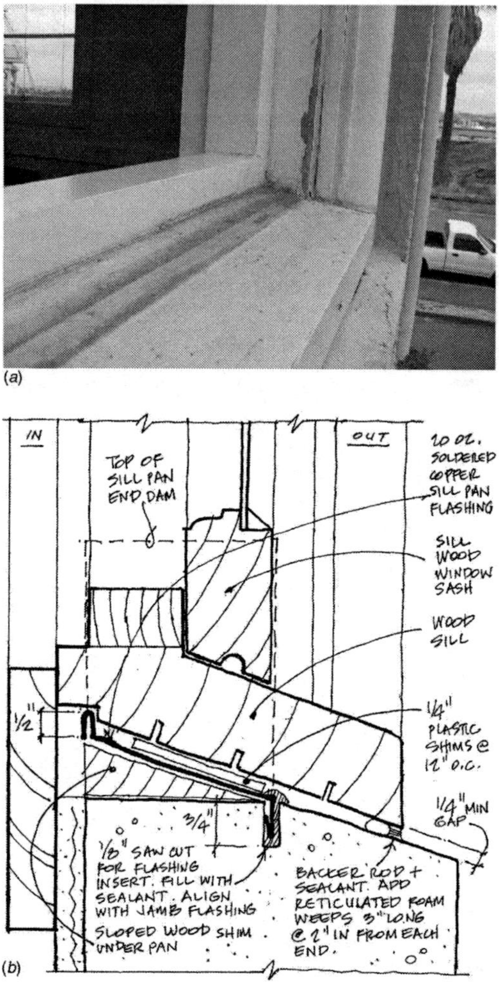

FIG. 13—*(a) Window sill. (b) Detail B. Sill flashing.*

Details

Supplemental metal flashing "fins" were added to the window frame installation in a new groove saw cut in the concrete wall opening. A sill pan flashing under the wood windows was also included. Both flashing features were concealed from exterior view behind the perimeter frame-to-concrete opening sealant joint. Reticulated foam weeps were incorporated with the sill sealant joint to permit sill pan drainage directly to the exterior (see Fig. 15).

The repair details included the drainage enhancing features of the existing recessed opening. The wall openings had existing sloped sills [5]. The jamb and head opening perimeter was uniform and appropriate for application of a uniform sealant joint and a continuous saw cut groove. The head, jamb, and sill were in a recessed opening that allowed water draining down the exterior wall to flow around the opening and drip away from the window [1,2]. See Figs. 12–15, details A–F. The rear leg of the new copper sill pan flashing created a continuous back-up dam to the exterior sealant joint for the wall opening.

(*a*)

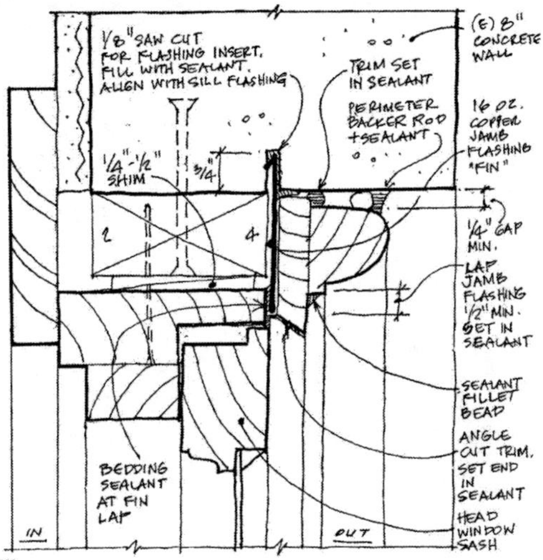

(*b*)

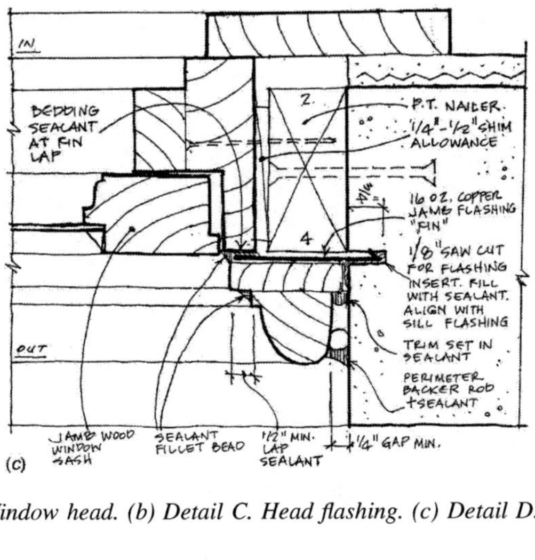

(*c*)

FIG. 14—*(a) Window head. (b) Detail C. Head flashing. (c) Detail D. Jamb flashing.*

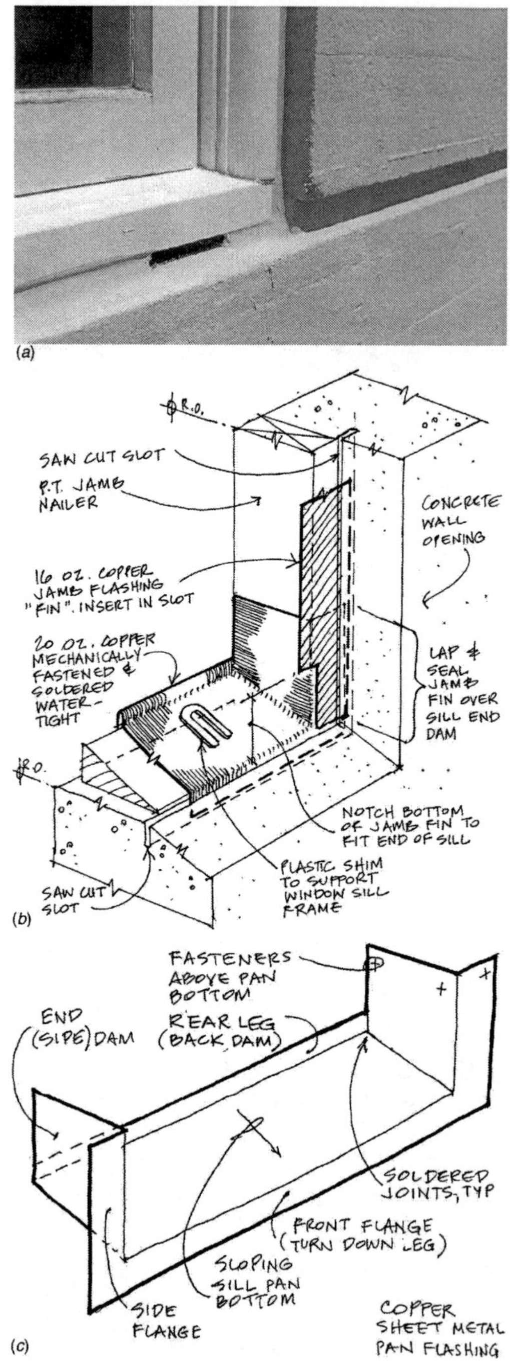

FIG. 15—*(a) Sill-jamb corner. (b) Detail E. Sill-jamb corner. (c) Detail F. Sill pan flashing.*

TABLE 1—*Comparison of Materials and Installation for Window Flashing.*

Materials	Alameda College-Case Study #1	Touro University-Case Study #2
Wall Construction	Poured-in-place concrete	Poured-in-place concrete
Window frame material	Aluminum (storefront), Blomberg Windows, Sacramento, CA. Bronze-Anodized Series 720, fixed sash.	Wood frame and sash, primed, painted, Ocean Sash & Glass Co., San Francisco, CA. Fixed and awning sash.
Perimeter frame sealant	Silicone, Dow Corning 795 plus Dow 1200 primer for concrete	Polyurethane, Sonneborn NP-1
Sill flashing material	Exposed galvanized sheet metal, 24 gage, soldered joints. Primer: Ameron "Amerlock 400;" Paint: Pittsburgh Paints "Industrial Enamel," 100 % acrylic	Concealed copper, 16 oz.@ jamb fins and 20 oz.; @ sill pan flashing, soldered joints
Reticulated foam weep	Polyester polyurethane, 30 ppi, Crest Foam "Filtercrest, S-Grade"	Polyester polyurethane, 30 ppi, Crest Foam "Filtercrest, S-Grade"
Glazing type	1" IGU, Milgard Inc.	Single glazed
Sill coating	VIP, "Dur-O-Coat 8100″ elastomeric	100 % acrylic paint

Wall Opening Design

The as-built openings for both projects had the advantage of existing design features that improved performance by protecting the perimeter sealant joints from exposure [6]. The openings had recessed head and sills with a smooth surrounding exterior finish. The Vallejo project had sloped sills and the Alameda project had a cast-in drip at the head opening.

Materials

Silicone was used as the perimeter sealant between concrete and aluminum frames. A dark color was selected for the sealant that matched the frame color and thus could remain unpainted. The perimeter sealant used between concrete and wood window frames was polyurethane in order to be easily painted when cured. The wood windows used concealed copper flashing. The aluminum windows used exposed galvanized sheet metal flashing that was mechanically fastened and soldered to fit each opening (see Table 1 for a comparison of flashing materials used for both case studies).

Lessons Learned

After eight years of in-place service, we returned to the projects to visually evaluate the performance of the window openings. We had not received any complaints from the clients during this period. We learned that the Touro University maintenance staff was about to repaint the wood windows.

In these case studies, the contractors for both projects were diligent about field measuring the existing openings and fabricating the sill pans to closely fit in position. The sill pans were well coordinated with the windows during installation.

Aluminum Windows

The aluminum window frame perimeter sealant joints are performing after eight years of service. The aluminum windows do not appear to have received any maintenance except for perhaps occasional window washing. The foam weeps used for the 1998 repairs are still in place and functional but appear in brittle condition on the south and west exposures (see "Wood-frame Windows" below). The dark, exposed sealant around the pan end dam extension could have been taped off to present a neater sealant line. The reglets were cut into the concrete with a hand grinder resulting in the appearance of only fair workmanship. The use of a saw cut groove and a template for cutting would have resulted in cleaner joints (see case study #1, Fig. 9(*b*)).

FIG. 16—*Reticulated foam weep deterioration.*

The exposed galvanized sheet metal sill pan flashing painted with enamel has remained in excellent condition. There have been no visible signs of corrosion even though the project is a few miles from San Francisco Bay.

Wood-Frame Windows

After seven years, the foam weeps are still in place, but some exposed on the weather side of the south and west are brittle and fragile to touch, although functional. After follow-up research with various reticulated foam manufacturers, material made from polyester polyurethane has been described as more susceptible to degradation from heat and moisture exposure than polyether polyurethane foam. See Fig. 16 for the brittle foam pried out of the sill sealant joint with a pen.

Polyether type polyurethane foam has been recommended for weeps by foam manufacturers during our latest research to improve exposure durability. However, we now understand that both polyester and polyether polyurethane foams may degrade with UV exposure. We would currently recommend deeper sections of foam material be installed into the gap of the sill sealant. This would provide foam material extended deeper into the gap to be protected from direct sun exposure. Replacement foam weeps could occur with the anticipated maintenance of future perimeter sealant replacement.

In the seven to eight years of service, little or no maintenance has occurred for either project. Maintenance, as well as material durability, continues to be a critical aspect of successful performance of window installations [1]. The sealant joints around the wood window frames to the concrete opening perimeter continue to perform. However, the wood window frames and sash in Vallejo show wear and the effects of lack of painting. The wood windows now require repainting and the exterior glazing putty is failing. The putty requires replacement in some locations facing the south and west exposures. The windows on the north and east sides appear in good condition.

Examining the as-built conditions reinforces the importance of certain aspects of window design that need to be considered when specifying products. The new wood window profiles lack the projecting horizontal mullions on the upper sash of the original window profile. The absence of this design feature increases the exposure of the horizontal sash to allow more sun and rainwater contact at the meeting joint. In this case, increased maintenance resulted from the ineffective window component design.

There was no weather-stripping included with the new wood windows. Weather-stripping should also be considered as part of the future maintenance of the operable sash-to-frame joints.

Conclusions

Similar projects should consider the inclusion of a sill pan flashing for block-frame window installations. The two examples show that different types of block-frame windows with sill pans can be installed with limited affects to the building appearance. The sill pan flashing enhances the building performance by providing a waterproof back-up system to the exterior sealed joints at the opening. This enhancement should be an important consideration where routine maintenance or replacement of sealant joints is not

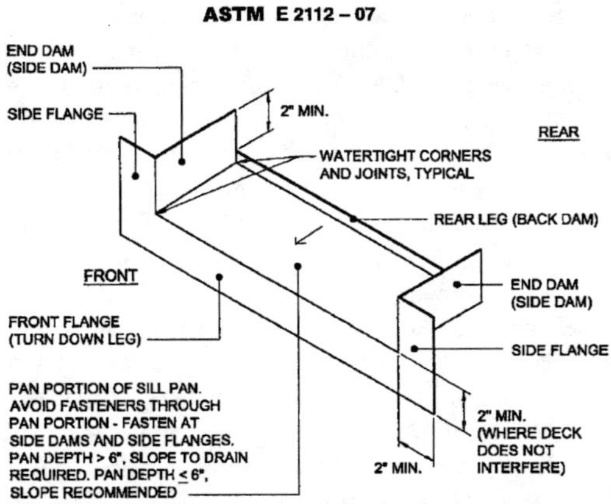

FIG. 17—*ASTM E2112-07 sill pan flashing configuration and dimensions.*

consistent or existent. Furthermore, the sill pan provides back-up to collect and drain water if there is a failure at the window frame sill corners in wood or aluminum frames.

The continuous through-wall sill pan flashing used at the Alameda project may not be suitable for projects in Northern or Southern climates of the United States, where thermal bridging could result in unwanted condensation.

Although the two case studies involved existing concrete walls, the detailing methods can also be applied for new building, window, and flashing design. The construction of poured-in-place concrete walls today is very similar to the walls constructed in the 1940s for Vallejo and the 1960s for Alameda. Block-frame windows are also relatively the same today, with respect to installation and the reliance on perimeter sealant joints. The inclusion of a sill pan flashing is not only currently good practice, but is also consistent with the recommendations of the latest ASTM window installation standards [3,5]. (See Fig. 17.)

Acknowledgments

The author wishes to recognize the contributions of Ken Klein and Michael Lardner at Simpson Gumpertz & Heger, Inc. for developing the initial window repair design for the College of Alameda Library.

[1] Ruggiero, S. S., and Myers, J. C., "Design and Construction of Watertight Exterior Building Walls," T. A. Schwartz, Ed., *Water in Exterior Building Walls: Problems and Solutions*, ASTM STP 1107, ASTM International, West Conshohocken, PA, 1991. Examples of sill pans include Fig. 13 with a sill pan in a concrete "punched" window opening with an upturned end dam sealed to the rough opening. STP 1107, Fig. 2 shows a sill pan supported by a sill clip with sloped front flanges sealed to the concrete and included with a drip edge. The perimeter opening is sealed with weeps at the sill joint.

[2] Myers, J. C., "Window Sill Flashings: The Why and How," *Prog. Arch.*, June 1990, pp. 41–43. Guidance includes sloped sill pans, weeps at sill joints, upturned rear leg sealed to window frame, and extended front of pan beyond the wall face with sealed front pan edge and drip. The sill pan with turned-up end dams should be sealed to the rough opening and the pan not fastened through the sill portion. Clip angles at the rear of the pan should be used. Pans and window frames should have tapered shims to provide slope for pan drainage.

[3] ASTM, Standard E2112-01, "Standard Practice for Installation of Exterior Windows, Doors, and Skylights," *ASTM Annual Book of Standards*, ASTM International, West Conshohocken, PA, 2001. See section 5.16—Sill pan flashings are recommended for sliding glass doors and block-frame windows. ASTM Fig. A3.2: Sill pans include rear leg, end dams side flanges and front flanges. Table A3.1 provides the height for the rear leg of pans to resist the rise of water due to different levels of wind pressure.

[4] *Architectural Sheet Metal Manual*, 5th ed., SMACNA, Chantilly, VA, 1993. See Fig. 4-21C for pan with slope. See Fig. 4-21D for typical dimension of sheet metal sill pan flashings. The end dam height and side flange are noted in SMACNA as 4 in. minimum. The rear leg is noted as 1/2 in. However, dimensions and fastening locations often have to be modified for specific conditions. By comparison, Case study #1 has an equivalent SMACNA side flange dimension. The end dam height matches the rear leg at 2 in. Case study #2 has a sloped sill and a 1/2 in. rear leg.

[5] ASTM, Standard E2112-07, "Standard Practice for Installation of Exterior Windows, Doors and Skylights," *ASTM Annual Book of Standards*, ASTM International, West Conshohocken, PA, 2007. In the revised ASTM Fig. A3.2, minimum dimensions are noted for the pan components see Fig. 17. See ASTM Sec. 5.16, which compiles and revises pan flashings information contained in the 2001 edition. Section 5.16.3 recommends pan flashings for all windows and doors with some exceptions noted. Table 4-A outlines four categories of pan flashing types by material and fabrication. Annex A3.1 briefly mentions the use of air seals to reduce the rear leg height of sill pans. The 2 in. minimum sill pan dimensions for sill flanges and end dams is less than the 4 in. from SMACNA. By comparison, Case studies #1 and #2 would be classified as Type I-Rigid Sheet, Multiple Pieces sill pans per E 2112-07. Case studies #1 and #2 use air seals at the interior opening perimeter to reduce the needed height of the sill pan rear leg. Case study #1 uses a support angle behind the pan to avoid fasteners through the pan.

[6] Lies, K. M., and Faith, B. A., "Window Detailing Considerations for Leakage Prevention," *Water Leakage Through Building Facades*, ASTM STP 1314, R. J. Kudder and J. L. Erdly, Eds., ASTM International, West Conshohocken, PA, 1998. Discusses secondary sill flashings intended to collect and evaporate water or to weep water out. Sill pans should slope, be attached at the sides or by a clip angle and not fastened through the pan. A continuous perimeter sealant should be included around the window and interior opening. By comparison, Case study #1 utilizes a clip angle, no sill fasteners, and a continuous air seal. Case study #2 includes a sloping sill and no sill fasteners. The interior trim casings act as an air seal.

Journal of ASTM International, Vol. 5, No. 10
Paper ID JAI100885
Available online at www.astm.org

Jared B. Lawrence[1] *and Paul G. Johnson*[2]

Envelope Remediation—A Case Study in Support of an Over-Cladding Approach

ABSTRACT: This paper addresses a case study in which a contemporary, high-performance glass curtain wall and a composite metal panel cladding system were installed over an existing failed 1960s masonry wall with aluminum windows. In addition to consideration of the typical performance requirements for air and water infiltration and thermal insulation, the new wall systems were designed to meet glass shard retention requirements and to accommodate improvements to the structural frame required by the General Services Administration. These performance requirements were all met while creating a more inviting exterior appearance, improving the interior work environment, extending the building's service life, and reducing energy consumption. Problems of the original design and construction, extent of previous remediation attempts, findings of the investigation of the existing building, and the development of remediation options are discussed to provide background to selection of the over-cladding approach for remediation. Opportunities and limitations posed by the existing building, LEED requirements, glass shard retention, construction strengthening, full and uninterrupted occupancy, as well as other challenges inherent in the over-cladding of an existing high-rise building on a tight urban site are identified and discussed, and positive results are identified.

KEYWORDS: over-cladding, performance, existing exterior envelope, unitized, panelized, curtain wall

Introduction

The vast majority of our existing building stock in the United States was constructed in the 20th century with the largest amount certainly constructed in the last half of the 20th century. When considered from a historic perspective you will find that this time period covers an extreme range of construction materials, procedures, design preferences, and performance expectations. For example, in 1900 there was no air conditioning, the use of electricity in buildings was very limited, horses and mules were common on construction sites, and material standards and testing were relatively new sciences [1]. ASTM was a new organization (founded in 1898) at the beginning of this time period with one of their first standards, "Structural Steel for Bridges," published in 1901 [1]. This was closely followed by the United States Government establishing the National Bureau of Standards in 1901. At the beginning of the 20th century, and for the most part through the first half of the century, technology and construction practices changed relatively slowly and in great part were driven by time proven technology, practices, and materials.

As time progressed, the use of modern machinery on construction sites increased; our society started to expect more and more in the way of creature comforts from our buildings; and modern science and manufacturing capabilities began to increase our options for materials. At the same time, our design preferences started to change and evolve. Interestingly, these changes in construction options, opportunities, and desires began to occur across increasingly shorter time periods. At times today we seem to be members of the "system of the month club." This is an important issue because in the past 50 years the rate of change in construction technology, practices, and materials has been driven by opportunity and desires, not necessarily the proven results and sound technology more typically seen in the earlier part of the century. The current approach in construction is often fed by cost concerns, "design" preferences, schedules, and worst of all, very good marketing. "Unfortunately, contemporary builders often ignore exemplars in a continuing but failed effort to fashion new images" [2]. All too often construction systems are not selected on the basis of sound professional judgment based on performance history, comprehensive ma-

Manuscript received November 8, 2006; accepted for publication September 15, 2008; published online November 2008.
[1] Associate, SMITHGROUP Building Technology Studio, Suite 1700, 500 Griswold, Detroit, MI.
[2] Principal, SMITHGROUP Building Technology Studio, Suite 1700, 500 Griswold, Detroit, MI.

terials or systems testing, or a thorough analysis and understanding of the materials or systems abilities and compatibilities. Instead, new materials, systems, and technology are presented, and accepted, as the answer to our problems without sufficient research and application history. When you add that time and exposure to the elements are the greatest enemies of *any* material and that our society demands more and more in the way of creature comforts and control of the interior environment, the result has been predictable; many of our existing buildings have exterior wall systems that will not, and cannot function as desired or as needed to keep them viable.

We currently have a large number of buildings in this country which have perfectly sound structures and other basic building systems that are clad with an exterior envelope that does not perform to current expectations. In many cases this inadequate performance is not due to the exposure conditions of exterior walls, the exterior design conditions have not changed, it is driven by the desire for improved performance and efficiency. This means that much of our existing building stock does not meet our current expectations and demands for comfort and thermal performance. This trend to higher performance buildings is not likely to stop in the future. Existing buildings are likely to fail our expectations in at least one of the following areas:

- Thermal efficiency.
- Lateral load resistance (wind).
- Acoustic isolation.
- Blast and impact resistance.
- Fire resistance.
- Water infiltration resistance.
- Air infiltration and exfiltration resistance.
- Seismic resistance.
- Self control technology and interaction (smart building systems).

Further, there is a growing desire to provide interior humidification to increase occupant comfort. This trend will only increase the pressure to find cost effective ways to both restore exterior envelope performance of existing buildings and to provide performance for considerations which the original construction was never designed to accommodate. Current building codes in the United States do not require buildings to be upgraded to meet contemporary performance standards in most cases. However, it is reasonable to assume that this may change in the future. Even under current codes, buildings often undergo extensive renovations to improve the overall performance of the envelope, improve or update the appearance, and to improve specific performance criteria that may be related to a change in use (occupancy) of the facility, or a combination there of.

This paper describes a project in which an over-cladding approach was utilized to successfully upgrade a building to current expectations for performance and appearance. Figure 1 [3] shows the building as originally designed and constructed; Fig. 2 shows the building upon completion of the over-cladding process. In many ways this project is an excellent example of many of the unsatisfactory conditions previously described and demonstrates an effective approach to resolving those problems. We will provide a general description of the over-cladding process, and identify:

- A brief history of the building.
- Problems it was experiencing.
- Concerns and requirements of the owner.
- Investigative process and identification of deficiencies.
- Evaluation of remediation approaches.
- Results of this project.

We will also address both the benefits and risks of the over-cladding approach, and identify the key elements that led to the success of this project.

The building's roof system, plaza waterproofing, and below grade waterproofing systems will not be addressed in this paper. Although these systems were replaced or remediated to various degrees, the corrective actions were relatively straight forward and other than budgetary considerations did not directly impact the evaluation, design, or corrective process for the exterior wall systems.

FIG. 1—*The original building.*

Over-Cladding

Concept

Over-cladding, as used in this paper, is a term describing the specific process of resolving exterior wall problems by adding a new exterior skin, or envelope, over the existing construction. This process will often require removal of portions of the existing wall construction, such as windows or curtain wall, prior to completion of the work. Depending upon the original construction it may also require penetration of the existing exterior systems in order to connect the new structural support system that transfers the lateral loads and the additional dead load of the new over-cladding systems to the frame of the building.

The primary advantages of over-cladding are:

- The ability to limit construction-related disruption of building occupancy.

FIG. 2—*The building after over-cladding.*

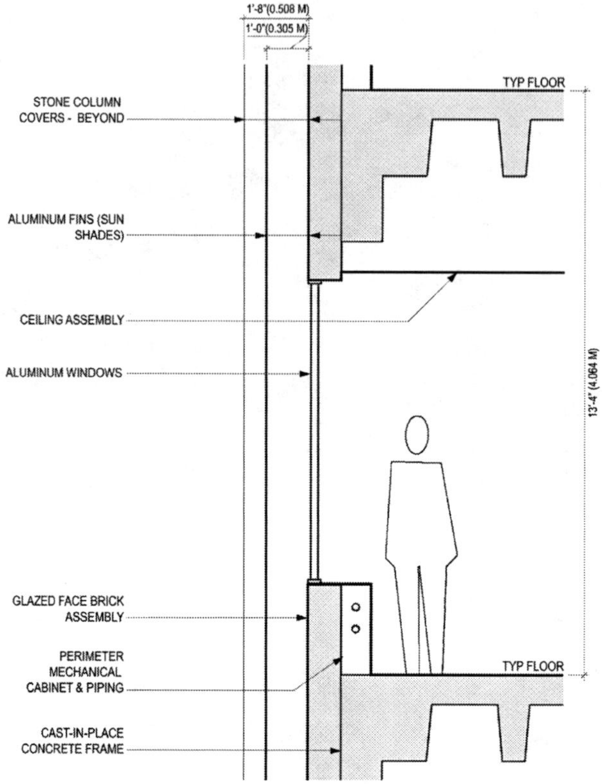

FIG. 3—*Section—original construction.*

- The ease of revising the exterior appearance, if desired.
- The opportunity to improve various aspects of performance such as blast resistance, thermal performance, and acoustic performance.
- The opportunity to stabilize existing deficient exterior envelope components or systems (leave in-place) without risk to safety or building performance.
- Much of the existing exterior can be retained and is not hauled to landfills, compared to removal and replacement of the entire exterior envelope.
- Reduced cost when compared to removal and replacement of the entire exterior envelope.

Disruption to building occupancy is limited by installing the new structural supporting system and the new weatherproof cladding before removing components of the original cladding system. This approach does not expose the building interior to the elements, thereby saving energy and cost used for temporary protection and heating necessary with typical construction sequencing and methods. This approach also reduces the interior disruption to removal of existing window or curtain wall components (e.g., glass and aluminum), and removal and installation of interior trim components.

The four basic steps of the process as described in this case study are shown schematically in Figs. 3–6. The figures illustrate the major steps associated with the over-cladding approach as applied, in section, to a typical floor at the north elevation of the building used as an example in this paper.

- The original building, (Figs. 1 and 3).
- Preparation for over-cladding: remove existing aluminum sun shade "fins," granite column covers, remove localized face brick, paint existing exterior, and install new structural girts, (Fig. 4).
- Install new cladding system: install the new glass and aluminum curtain wall system from the

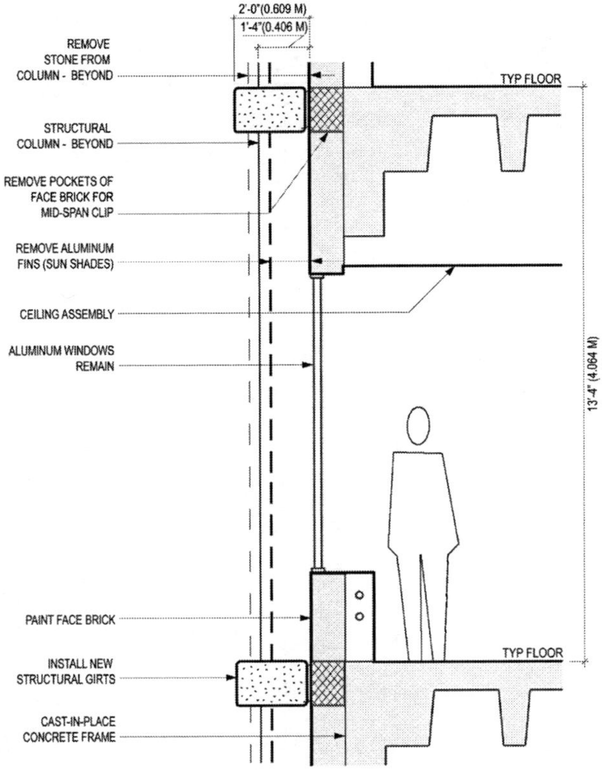

FIG. 4—*Section—preparation for over-cladding.*

outside, maintain the existing windows "as-is" to provide protection to the interior finishes during erection of the curtain wall, (Fig. 5).

- Remove windows and install trim: remove existing windows, paint exposed structure (girts), and install new trim at the existing window opening and columns, (Fig. 6).

This approach is quite straightforward, especially when considering the associated benefits compared to risks, and it can easily be applied to a wide variety of building systems. The benefits and risks of over-cladding are also straightforward for evaluation purposes.

Benefits and Risks

Over-cladding as extensive as this case study requires a comprehensive evaluation of the existing conditions to understand the risks and benefits.

A summary of benefits identified during the design process for this project included:

- Less disruption to current occupancy than other remediation approaches.
- Significantly less solid waste due to limited material removed.
- Construction time in the field could be reduced.
- Work required to prepare the building for the new cladding system could proceed with minimal disruption to occupants.
- The relative ease of improving thermal performance for the entire building envelope.
- The ability to dramatically alter and improve the building appearance.
- The ability to dramatically improve overall building envelope performance.
- Glass shard retention capabilities could be efficiently incorporated in the new glazing assemblies

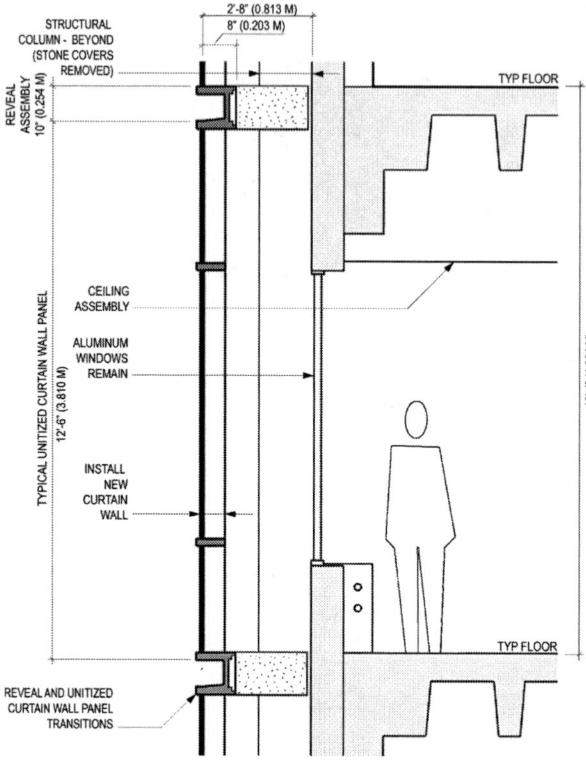

FIG. 5—*Section—install new cladding system.*

when compared to the cost and effort associated with retrofit application to attain the same performance.

A summary of potential risks identified for this project included:

- The initial cost was anticipated to be higher than the cost for a less comprehensive remedial action.
- The presence of hidden structural flaws would not be identified in areas where the existing system(s) are not removed for installation of the new cladding system.

As in all decisions related to design and construction, the appropriateness of this approach will depend upon the specific goals, opportunities, and constraints of the project. The over-cladding approach is probably best suited to situations where significant changes in appearance and performance are desired, as in the case study.

Case Study

Building History

The Des Moines Federal Building is a ten-story multi-tenant office building in the heart of Des Moines, Iowa. The building, with an occupancy rate of 97 % at the time, provided office space for 42 different United States Government agencies. With approximately 850 full time occupants, ranging from senators to FDA field agents, plus a significant number of public visitors every day, the building was anything but a standard construction site. In 1999 SmithGroup (the owner's consultant) was commissioned by the build-

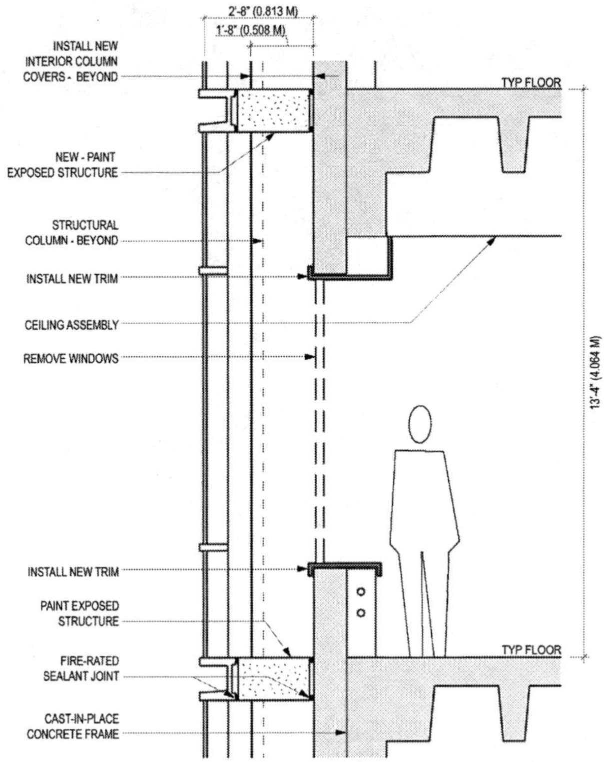

FIG. 6—*Section—remove windows and install trim.*

ing owner (the General Services Administration) to determine the cause(s) of water infiltration through the exterior walls, identify remediation opportunities for the leakage, and recommend a construction budget for review.

Prior to this project several attempts had been made to correct air and water leakage; however, they were only partially successful. These efforts consisted primarily of flashing repairs and, interestingly enough, permanently sealing all operable windows in the closed position. These previous attempts appeared to have been based upon a "piece-meal" approach, attempting to solve what appeared to be individual problems as opposed to a comprehensive approach that considered the materials, systems, and assemblies used in the walls, and their interrelationships. The result was that much of the disruptive and damaging air and water leakage was not corrected, and the problems continued as demonstrated by water leakage, (Fig. 7). This owner-provided photograph illustrates the extent of water leakage experienced at the head of a typical window condition during a rain event prior to implementation of the over-cladding project.

As indicated, the occupancy rate of this building was very high with no capability to relocate tenants to "swing" spaces within the facility during the construction process. Moving tenants out of the building on a temporary basis was considered to be unacceptable and limiting disruption to the tenants and the public using the building was of primary importance.

The Problem

Constructed in the 1960s, the building was not thermally efficient, had developed severe and disruptive water leakage problems, and allowed unacceptable levels of air infiltration and exfiltration. The exterior masonry had systemic cracking, the flashings were not working properly, and anchorage was found to be

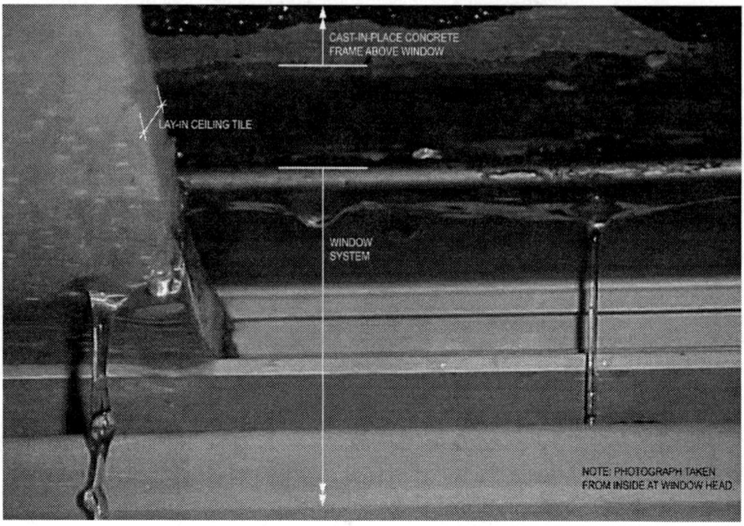

FIG. 7—*Wall leakage in 2000.*

questionable. The exterior granite cladding of the building's perimeter columns were also in poor physical condition and potentially unstable due to failure of some stone at anchor locations, a condition that led to a stabilization program prior to award of the over-cladding project to the design/build contractor (DB). In addition to these physical deficiencies and problems, the owner was not satisfied with the public image of the building; a new, more inviting appearance was desired. Further, due to the bombing of the Murrow Federal Building in Oklahoma City in 1995 [4], the Federal Government mandated changes to selected federal buildings to improve resistance to glass shard release and progressive collapse of a building's structural frame in the event that a facility was subjected to a blast event. The original structural system and exterior wall assemblies of the facility did not provide an acceptable level of performance.

The need to correct existing performance deficiencies plus the overall condition of the exterior envelope provided an ideal opportunity to also update the building's aesthetics (change its perception by the public), increase energy efficiency, improve occupant comfort, and bring the building's exterior envelope into compliance with mandated owner performance requirements.

The Investigative Process

The investigation and evaluation process was started in the fall of 1999. This effort would eventually entail:

- Review of available original construction documents.
- Review of previous leakage investigation reports and repair attempts.
- Review of consultant's blast threat analysis.
- Observation of existing conditions, including intrusive probes and testing.
- Evaluation of the findings.
- Analysis of failure modes.
- Development of concepts to correct the existing deficiencies.
- Development of concept budgets to aid the owner's planning and evaluation process.
- Preparation and presentation of a report of findings and recommendations.

The Des Moines Federal Building was originally constructed with:

- Cast-in-place concrete frame.
- Walls of soldier course glazed face brick laid in stack bond.
- Concrete masonry unit or brick back-up (varied by location).

FIG. 8—*Removal of vertical sunshade "fins."*

- Windows of single-pane glass with extruded aluminum frames; windows from the second floor up were originally operable.
- Granite cladding of exposed columns.
- The mechanical system was fed around the perimeter of the floors, concealed in metal cabinets below the windows. The cabinets housed the forced-air and fin tube components.
- Heavy aluminum vertical sunshade "fins," (Fig. 8).

Generally the building had very little thermal insulation, and what insulation it did have was only in localized areas. The mechanical and electrical systems were typical to the time of construction and the region.

The owner reported that the exterior walls had been problematic since the original construction and that despite repeated and considerable efforts to resolve the conditions only limited success had been achieved. Previous attempts by the owner to repair the building's excessive air and water infiltration problems included sealing the vent windows shut, replacing masonry flashing at the shelf angles of every floor, and localized (minimal) repair to the displaced granite column cladding. Unfortunately, this work was completed without a comprehensive evaluation of the building's envelope and therefore without a full understanding of failure mechanisms inherent to the building's original design and construction. The significant effort and resources expended in these attempts resulted in minimal improvement to the performance of the building's exterior envelope.

The investigation by the owner's consultant of the masonry walls identified the following deficiencies:

- Improper and discontinuous horizontal wire reinforcing anchorage of the face brick.
- Cracking of both mortar and facing brick.
- Deteriorated interior finishes from moisture infiltration.
- No masonry control or expansion joints, resulting in the displacement of face brick at corner conditions, (Fig. 9).
- Inconsistent anchorage of the face brick.
- Deficient flashings.

Investigation of the windows identified the following deficiencies:

- Framing and sash were not thermally improved.
- Single-pane glass was not energy efficient.
- Weather stripping and gaskets were not functional (water and air infiltration).
- All operable sashes were sealed shut and no longer operable.

Investigation of the granite column cladding identified the following deficiencies:

- Sealant joints between panels were improperly installed and deteriorated.

FIG. 9—*Existing face brick.*

- Some stone panels were spalled and no longer properly engaged at anchor points.
- Granite panels were cracked.
- Water entering the granite cladding had corroded the unprotected steel anchors embedded in the concrete frame.
- Water entering the granite cladding was accumulating in loose fill insulation and had consolidated the insulation at each floor resulting in reduced insulation value.
- The consolidated insulation would accumulate water and freeze, displacing granite panels and compromising gravity support, (Fig. 10).
- Flashing was not provided at shelf angles to manage water infiltration.
- Lateral load anchors were inconsistent in placement.

Note that there was no in situ testing of the existing exterior envelope to quantify air and water leakage or infiltration, and there was no testing of existing thermal performance.

Conclusions and Direction

Based on these findings the owner and their consultant determined that repair of the specific deficiencies identified would not be cost effective. Many of these deficiencies were inherent to the design of the systems and construction methods used. Repairing these problems in-place (the traditional approach) would also impose a level of disruption to the building occupants that was unacceptable to the owner. With these considerations in mind it was recommended to keep the existing envelope in place and over-clad the building with a new curtain wall system. This would allow for correction of not only the identified inherent deficiencies but would also accomplish significant improvement to the energy performance of the building by use of contemporary high performance curtain wall systems and the addition of thermal insulation By providing a completely new envelope system the service life of the building's exterior would essentially restart. The approach could also be accomplished quickly and efficiently, essentially from the outside, with minimal disruption to the occupants.

An additional benefit of this approach is that it allowed stabilization of the existing masonry without a separate cost and effort to access the building exterior for removal and repair. By over-cladding, the existing systems were protected from further deterioration due to exposure to the harsh Iowa climate. The over-cladding also protected the existing masonry from further exposure to lateral (wind) loading. Prior to reaching this decision, areas of masonry with serious concerns (Fig. 9, for example) were evaluated to determine if stabilization or repair was required prior to over-cladding. This evaluation included consid-

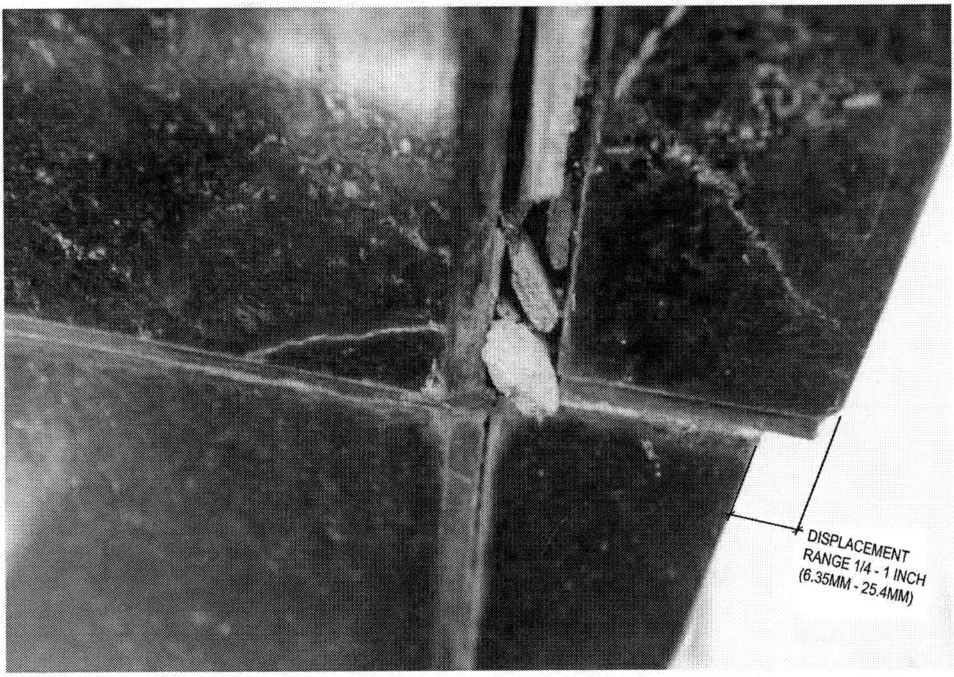

FIG. 10—*Displaced stone column covers.*

eration of the age of the brick when compared to the theoretical amount of irreversible growth due to moisture. Due to the age of the brick, it was determined that the vast majority of the irreversible growth had already occurred; therefore, additional stress due to growth would likely be insignificant [5].

Based on these recommendations the owner decided to proceed with development of an over-cladding design for the building that would:

- Remediate the identified deficiencies.
- Accommodate the long-range goal of achieving a LEED Silver rating for the building.
- Incorporate glass-shard retention capabilities to the glazed panels and resistance to progressive collapse of the structural frame to meet the prevailing owner standards.
- Maintain full occupancy of the building during construction.
- Provide a new building aesthetic based on the owner's current philosophy of design.

The Design Process

Design and Aesthetic Considerations—Building aesthetics is not the focus of this paper. It is, however, important to note that the over-cladding process, as identified in this case study, provides the opportunity to significantly improve the appearance of a building, while correcting technical deficiencies and enhancing physical performance.

In this case, the over-cladding system was designed based on the requirements of the owner's Design Excellence Program [6]. Under this program, the design undergoes an extensive review process to assure that it meets a high standard of aesthetic integrity by incorporating input from distinguished private-sector design professionals (appointed national peers by the Commissioner of the Public Buildings Service) and various owner personnel. The review is conducted in three stages during the schematic design process. Starting from multiple conceptual designs the approach was reduced to the final design, which was radically different from the building's original design. The building's rigid and imposing symmetrical elevations, dark tones, repetitive lines, and awkward proportions were replaced by a contemporary design that presents a more relaxed, comfortable, and transparent façade to the community. The types, opacity, proportions, and color of the materials used in the over-cladding system address differences in the use and orientation of each elevation. The highly-transparent north elevation is used to establish a natural orientation for the building and identify the public side that is utilized daily by visitors.

The north elevation, which serves as the main entrance for the public, was over-clad with a clear glass curtain wall system that spans from the second floor to the top of the parapet at the roof nine floors above. The curtain wall system utilized high performance vision-glass and spandrel panels and a metal reveal to conceal the edge of the new structural girts at the existing floor lines (Fig. 5). The system extends the length and height of the building's elevation and is terminated 2 ft 6 in. (0.762 m) past the first and last vertical mullions creating glass "wings" that help identify the system as a plane independent of the building frame and adjacent elevations; this also adds to the apparent lightness of the elevation. The south elevation, predominated by service uses and the employee entrance, was over-clad with a simple and efficient curtain wall system comprised of high-performance clear glass and a composite metal panel glazed into the aluminum mullion system at the spandrel panel. This system helps to reduce the thermal load of the wall through the use of light colors, and provides a durable exterior envelope system with clean lines, increased thermally efficiency, and minimal maintenance.

Performance Requirements—General wall performance requirements, including air, water, and wind resistance were based on the International Building Code at a minimum. However, most of the materials and performance levels provided for various systems exceed the minimum established by code and are representative of high quality institutional construction. Performance requirements for allowable air and water infiltration of the glazed curtain wall and metal panel systems was established using ASTM and AAMA standards with values determined appropriate by the design team. Preconstruction testing was performed for both systems to verify the design would perform as anticipated. These tests were conducted by an independent testing agency. The curtain wall system was tested at the testing agency's facility while the metal panel system was tested using mock-up panels that were assembled at the contractor's yard. In situ testing was also conducted, utilizing portions of the systems installed on the building, by the testing agency utilizing AAMA 501.2 at several locations to verify acceptable installation of the components and transitions between systems. These verification tests were effective in identifying deficient conditions of the installed assemblies; appropriate modifications were implemented to assure compliance with the performance requirements.

As previously indicated, the owner did not require a specific level of thermal performance in the program of renovation for this building. However, the use of contemporary high performance exterior wall materials and systems greatly improved the thermal envelope. Repair and replacement of significant portions of the mechanical and electrical systems of the building subsequent to this project made a simple and straightforward analysis and comparison of energy usage impractical following completion of the over-cladding process.

There were also several performance requirements specific to the owner and federal building standards. These requirements proved to be most challenging to incorporate. However, the over-cladding design was developed to incorporate these special requirements and provide a fully integrated solution:

- Glass Shard Retention–A specific level of protection was required for the danger presented by glass fragments that may become airborne during a blast event. This requirement is selected based upon a category and location of blast, and a desired level of protection against glass shard release, and their theoretical travel distance. The threat level and specific design requirements were determined by the owner through a separate threat assessment prior to the inception of the project. A blast consultant, Applied Research Associates, was retained to assist in the development and evaluation of systems and materials to meet the design requirements and to ensure the solution was appropriate and viable. They also assisted in the review of shop drawings and the blast and structural calculations for the curtain wall and strip window systems provided by the design/build contractor.
- Construction Strengthening–Late in the design process, with the basic systems and aesthetic design decisions essentially complete, the owner requested an analysis of the building with regard to progressive collapse [4] considerations based on a required charge size (explosive force) and an established stand-off distance from the building. This analysis, and subsequent design, was to assume the loss of specific key structural elements of the building and identify strengthening approaches to the building frame that would prevent progressive collapse of the remaining building

frame. This analysis was completed by the owner's consultant. Upon completion of a comprehensive analysis of the building's structural frame and preparation of alternative conceptual design concepts that would provide the desired performance for the frame, options were presented to the owner. Based on the review of the analysis and alternatives, the owner selected a specific concept for implementation. The system was eventually refined and a supplemental structural system was developed and coordinated to fit within the geometry and confines of the new over-cladding systems.

- LEED–"The Leadership in Energy and Environmental Design (LEED) Green Building Rating System™" [7] is the nationally accepted benchmark for the design, construction, and operation of high performance green buildings. Through recognition of increased performance in five key areas of human and environmental health the LEED program provides a guideline for measuring and documenting key aspects of sustainability:
  - Sustainable site development.
  - Water savings.
  - Energy efficiency.
  - Material selection.
  - Indoor environmental quality.

For this project, only the last three aspects of LEED (above) were considered due to the limited scope (i.e., no mechanical or electrical renovation). The owner understood the completed over-cladding project would not meet the requirements for the LEED program being limited to partial renovation of an existing facility; however, they did require this project to achieve as many credits as were feasible within the scope of work. This requirement was established to allow the future achievement of a LEED Silver rating upon completion of other long-range renovations that would eventually include significant work to the electrical and mechanical systems. It also required that whenever technically feasible, materials removed from the building were to be recycled. These requirements added another level of complexity and consideration to the selection and design of certain systems. The effectiveness a recycling program can have in the reduction of construction debris normally sent to landfills is illustrated by the quantity of materials recycled during this project [8]:

- Aluminum fins—95.5 tons
- Aluminum window framing—35 tons
- Granite column cladding—550 tons
- Concrete paving—36 tons
- Concrete masonry—82 tons

In the true spirit of recycling, the DB contractor even reused the original aluminum fins in lieu of traditional framing in areas where temporary protection was required during construction. These fins were only sent to the recycling facility after they had provided this last service to the building.

Key aspects of this project that directly relate to the LEED program were:

- Recycling of materials removed from the building.
- Use of materials manufactured within 500 miles (805 km) of the site.
- Selection of materials and systems with high thermal efficiency, low maintenance, and long proven life expectancy.
- Improvement of the indoor environmental quality through extensive use of glazing for daylighting, reduced air infiltration, and increased thermal efficiency for increased occupant comfort.
- Maximized use of existing building frame and other construction while minimizing material removed from the building that might otherwise be sent to landfills.

The final impact of the owner's decision to require LEED compliance will not be known until the remainder of the building renovation work is completed. However, looking at material and labor saved by reusing the existing construction and the list of recycled materials clearly indicates success. An over-cladding approach can, and in this case did, result in significant value in the area of environmentally responsive design and construction practices.

Material, Systems, and Installation—The new cladding system was designed to be supported by a new structural girt system that was attached to the existing concrete frame on each elevation. This system was installed outside of the existing building envelope (Figs. 4 and 11). The new girt system was comprised of structural steel tubes on the south, east, and west elevations to minimize the distance between the existing

FIG. 11—*Girts and unitized curtain wall on north elevation.*

and new exterior surfaces while precast concrete girts were utilized on the north elevation where the new cladding system was held farther off the existing frame. Precast concrete was utilized on the north elevation, at the request of the DB contractor, in order to efficiently extend the fire-rated floor assembly to the new exterior wall line, expedite the finishing of the girt, and to minimize disruption to the occupants.

The north elevation cladding consists of a combination of a manufacturer's standard unitized curtain wall system (Fig. 11), installed in panels with nominal dimensions of 5 ft (1.529 m) wide by 12 ft 6 in. (3.81 m) tall. The system utilized a manufacturer's standard unitized structural silicone glazed system with a 10 in. tall (0.254 m) custom component at the floor line to provide a deep reveal that creates a shadow line to articulate the building floor-to-floor module.

The custom steel stud panelized system utilized 4 mm composite metal panels (with a fire-resistant core) installed onto building wrap, rigid insulation, and cold-formed steel studs in the contractor's shop. A nominal 8 in. thick panel was fabricated as large as 30 ft (9.144 m) long by 13 ft 4 in. tall (4.064 m) allowing for quick and efficient installation from the outside, with a high level of quality control. For example, a large percentage of the primary air and water resistance sealant joints (panelized steel stud to panelized steel stud illustrated in Fig. 12) were installed under controlled conditions in the fabrication shop. The secondary barrier, comprised of commercial quality building wrap in the field of the panels with a continuous foam gasket around the perimeter of each panel, provided a secondary system of protection from air and water infiltration and was also installed in the fabrication shop (Fig. 12). A custom panelized steel stud and composite metal panel column cover, also a nominal 5 ft (1.524 m) wide by 13 ft 4 in. (4.064 m) tall was provided at the perimeter columns (Fig. 13). The systems installed on the east and west elevations are variations of the systems utilized on the south elevation.

The cold-formed steel panels were fabricated and stock-piled in a nearby warehouse acquired by the DB contractor for that purpose. Each panel was fitted with continuous "Z" sub-girts and extruded foam insulation between the sub-girts. The building wrap was installed to the exterior face of the sub-girts, and insulation was returned over the edge and taped to the stud assembly at the perimeter of each unitized panel. The composite metal cladding system was then installed to the sub-girts with self sealing tape used at the attachment clips where fasteners penetrate the building wrap. The perimeter foam gasket tape was also installed before the panels were lifted onto shipping "A" frames and moved a few blocks to the project site. The panels were then unloaded to a small staging area adjacent to the building (Fig. 14). In order to meet the constraints of the site, the panels were shipped to the site on an as-needed basis to maximize site space and minimize the potential for damage prior to installation and then lifted into place (Fig. 15).

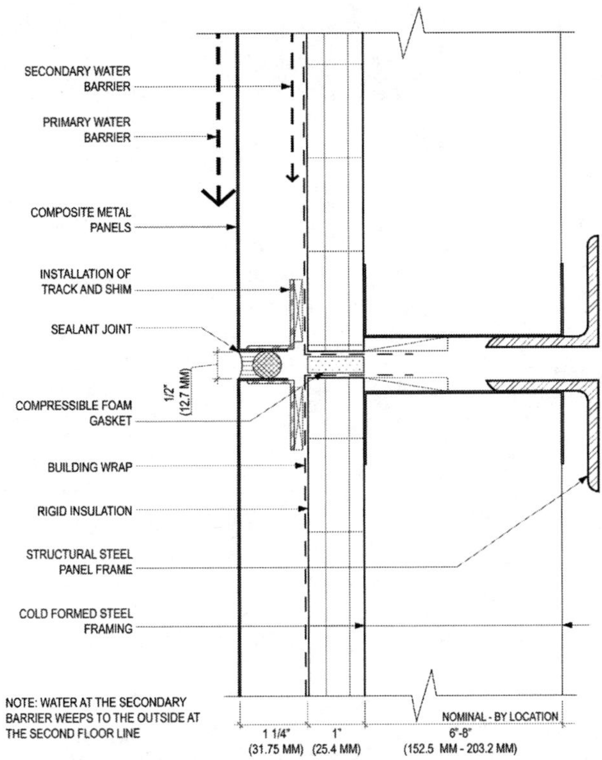

FIG. 12—*Section—typical water management for composite metal panels.*

Existing Concrete Frame—Significant difficulty was encountered when the steel erection contractor began installing the girt clips to the columns. It was soon determined that the clip detail would require significant modification to provide more dimensional latitude in order to work around the reinforcing steel in the existing concrete columns. Based on the column dimensions and the loading requirements, the new attachment clips for securing the new structural girts to the existing structural concrete frame consisted of a 3/8 in. (9.525 mm) thick bent steel plate, 3 ft $11\frac{1}{2}$ in. tall (1.207 m) by 1 ft 4 in. deep (0.406 m). This large size was required to allow for flexibility in locations of anchor bolts. This modification illustrates the importance of full consideration of the existing reinforcing steel, including the laps and rebar transitions between the floor plates and columns. When designing connections for an existing cast-in-place concrete building, flexibility and field adjustment is critical.

Procurement Process

This project was selected by the owner for a Design-Build procurement contract. Under this process an architect or engineer with special capabilities for the particular problem or need is selected as the design or "bridging" architect in the terms used by this owner. The consultant who performed the investigation was selected as the bridging architect for this over-cladding project. Design documents were developed for the building remediation to approximately 35 % completion, or the design development level. This design was then released to prequalified teams with each team consisting of a general contractor, appropriate design professionals, specialty contractors, and primary systems manufacturers to solicit Design/Build proposals. In this case the design professionals included an architect, a structural engineer, a curtain wall manufacturer, and a blast consultant. Based on this selection process Neumann Brothers of Des Moines, Iowa was selected as the DB contractor.

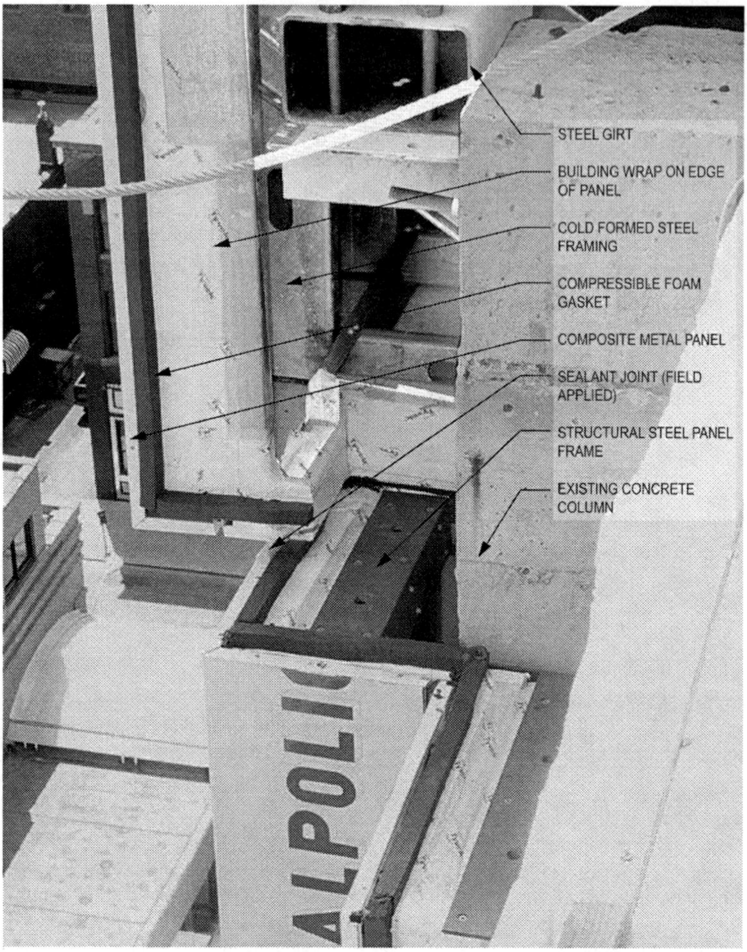

FIG. 13—*Metal panel column cover at south elevation.*

The owner's consultant assisted in the selection of the DB, and continued as a consultant to the owner throughout the remainder of the project as the contracting officer's representative.

Summary

This project accomplished the goals of the owner including:

- Improved performance.

  Eliminated water leakage, as advised post occupancy by the owner.

  Brought the exterior envelope into compliance with owner standards for air infiltration and exfiltration by the use of contemporary wall systems.

  Improved the security of the occupants and public through compliance with contemporary glass shard retention and construction strengthening requirements.

  Significantly increased the thermal efficiency of the exterior envelope by the addition of thermally efficient curtain wall and glazing systems, as well as added insulation to the opaque walls.

FIG. 14—*"A" frames being off-loaded at site.*

- Improved appearance.
- Resolved the problematic masonry and stone conditions.
- Complied with LEED goals.
- Minimized tenant disruption.

In addition to accomplishing the goals of the GSA, the project successfully demonstrated the potential for use of the over-cladding process in rejuvenating a nearly 40-year-old building and extending the facility's useful life into the future. All of this was accomplished in addition to a greatly improved appearance and performance. The improvements attained by this over-cladding approach were so successful that the building received a citation for modernization from the GSA 2006 Biennial Design Awards program.

FIG. 15—*Composite metal cladding panel being lifted into place at east elevation.*

While the over-cladding process is certainly not the only approach to be considered when facing the problem of building envelopes that can no longer provide appropriate levels of performance, do not provide the desired appearance, or are simply worn out, it is one approach that should be evaluated when deciding how to correct extensive problems with the various assemblies of the exterior envelope. The benefits, flexibility, and results can be dramatic.

References

[1] *1898–1998 A Century of Progress, Early Standards Development and the Origins of ASTM*, Chapter One: A Broader View, The Birth of Consensus, ASTM International, West Conshohocken, PA, 2006.

[2] McDonald, W. H., and Lewis, M. D., "The Importance of Studying Exemplars When Designing Stone Facades," *Performance of Exterior Building Walls, ASTM STP 1422*, P. G. Johnson, Ed., ASTM International, West Conshohocken, PA, 2003, p. 54.

[3] Photo provided by Heartland District of the General Service Administration, Property Management Division, Technical Support Branch (6PC), 1500 Bannister Road, Kansas City, MO.

[4] Interagency Security Committee, "ISC Security Design Criteria for New Federal Office Buildings and Major Renovation Projects," 2001.

[5] The Brick Industry Association, Technical Notes on Brick Construction 18 Revised: Movement Volume Changes and Effect of Movement Part 1, January 1991.

[6] U.S. General Services Administration, Design Excellence and the Arts, Program Goals and Features, GSA website, 2006.

[7] Leadership in Energy and Environmental Design (LEED), United States Green Building Council, USGBC website, 2006.

[8] Information provided by Neumann Brothers Construction, 1435 Ohio Street, Des Moines, IA 50314.

Journal of ASTM International, Vol. 4, No. 10
Paper ID JAI100863
Available online at www.astm.org

Jon M. Boyd[1]

Over-Cladding of Aluminum Framed Curtainwalls and Skylights

ABSTRACT: The need to upgrade the performance of existing building façade systems provides many opportunities for innovation and the creative reuse of existing building components. As many mid-20th century building enclosure systems age, they exhibit performance deficiencies, both in the form of deteriorating performance and in lagging performance as weighed against newer technologies. The installation of replacement exterior envelope systems on existing facilities provides many challenges, especially in fully occupied buildings. Over-cladding of existing metal and glass curtainwall and skylight systems, wherein portions of the pre-existing systems are integrated into the new construction, provides significant opportunities to resolve construction phase issues while providing improved functionality to the completed building envelope.This paper will address two case studies in which contemporary aluminum and glass elements were combined with pre-existing mid-1900s aluminum framed cladding elements to provide two wholly new enclosure systems with significantly improved performance characteristics. The case studies deal with two very different sets of project objectives, site constraints and project solutions, but share common approaches and eventual solutions. The paper presents the problems posed by the pre-existing systems, the site investigations conducted to confirm existing conditions and site restrictions, reasons for considering the specific replacement cladding approaches, specific project constraints, the approaches used in development of the new over-clad solutions, and considerations in the design, development, and construction phase delivery of these projects. Applicability of these approaches to other types of contemporary building enclosure types is explored as well.

KEYWORDS: curtainwall, skylight, over-cladding

Introduction

Modern era metal and glass curtainwall and skylight systems have evolved from manufacturer driven design concepts focused on maximizing economies in fabrication and erection. Early designs of both steel and aluminum types were devoid of the thermal properties and characteristics common in most well engineered wall systems of today. Pre-1970s glazing was typically single pane glass prone to significant condensation issues, often leading to water damage that frequently led to distress of other components and generally poor performance of the walls. Many early designs relied heavily on degradable exterior sealants and weathering gasket systems, and often had ineffective or insufficient water control mechanisms.

In contrast, newer, more energy conscious curtainwall and skylight system designs provide significantly better performance through the introduction of thermal improvements throughout the metal components, isolation of interior and exterior components, the use of energy saving insulated glass units, specialty glass coatings, and improved insulation and vapor control materials. Most new cladding systems today typically incorporate superior sealant technologies, rain-screen and pressure equalization approaches, and superior moisture control mechanisms to systems common in the 1950s through the early 1970s.

As our building stock continues to age, many owners will look to upgrade existing curtainwall and skylight systems to address a variety of concerns. These can range from systems that simply don't keep the rain out, to more subtle but objectionable performance deficiencies. Others will simply want to put new life back into their buildings, update the aesthetic, or raise a class "B" or "C" building to compete in the marketplace.

In many of these instances, over-cladding of existing exterior enclosure systems may prove to be economical, efficient, as well as environmentally friendly, while addressing significant challenges that

Manuscript received October 18, 2006; accepted for publication June 26, 2007; published online December 2007.
[1] Principal Architect/Structural Engineer, Executive Vice President, Klein and Hoffman, Inc. (K&H), 150 South Wacker Drive, Suite 1900, Chicago, IL60606.

occupied buildings face that new construction does not. The over-cladding approach addressed herein, applies to a group of cladding systems with mullion elements, primarily of the "stick-built" variety where linear structural elements provide the structural backbone of the system, as opposed to panel type systems. Two case studies illustrate how this approach has been used successfully to solve a variety of challenges.

Background

The over-cladding approach outlined is quite simple. The existing in-place mullion system is used as an armature upon which new structural or glazing components, or both, are applied in a "piggy-back" fashion. The new custom fabricated components provide the air and water control mechanisms, and where required, work in tandem with the original mullions to provide the necessary structural capacity for the new installation.

In many instances where recladding is performed, the newly installed systems are required to withstand higher structural loadings, including the added weight of insulated glass units and higher code mandated wind, snow drift, or seismic loadings than those in effect when the original constructions were put in place.

Case Study—1: Skylight Over-Clad

The first example of this specific over-cladding approach involves a highly articulated aluminum skylight system on a significant house of worship in the upper Midwest. The building, constructed in the mid-1960s, is a cast-in-place structure consisting of 16 curvilinear concrete forms which support a series of 14 finger-like skylight elements. The original skylight was state-of-the-art shop fabricated construction at the time it was built. The system consisted of aluminum "I" sections supporting single pane wire-glass, with exterior applied aluminum glazing caps. A lack of thermal separation and insulated glazing created considerable condensation related water problems early in the life of the structure. Thermal movement of the system, coupled with aging sealants, led to eventual water leakage from wind blown rains, and heavy melting snows. Several early attempts at remediation improved the precipitation related problems, but could not resolve the inherent condensation related deficiencies. It was clear that upgrading the system to include both thermal separation and insulated glazing was necessary to resolve the ongoing issues. Due to the architectural significance of the structure, the solution needed to be sensitive to the original design, and provide for minimal aesthetic change.

The original glass installed in the skylight was wire-glass, which had a characteristic gray appearance. Prior glass replacements utilized laminated heat-strengthened glass required by current code, which had a

FIG. 1—*A typical view of the exterior of the skylight prior to the over-cladding project is shown. Note the variation in appearance between the prior glass which had a green-blue hue and the wire-glass of the original construction. This variation gave a patchwork appearance to the skylight at the exterior, and created a varied quality of ambient light at the interior during daylight hours.*

FIG. 2—*The appearance of the sanctuary from the interior is illustrated and the interior lay-lights that are suspended from the outer skylight frame are shown. Eliminating the necessity to remove this installation during construction became a primary determinate leading to the over-clad approach for the project. Erecting scaffolding to provide access to the interior lay-lights would have added significant time and expense, which were avoided by using the construction methodology of an over-clad installation.*

green-blue hue. The replacement glass contrasted significantly with the original wire-glass, giving a patchwork appearance to the skylight at the exterior, and created a mix of ambient light color at the interior during daylight hours, see Fig. 1.

One of the project objectives included the desire to eliminate the necessity of removing the interior lay-lights that were suspended from the outer skylight framing; this became a primary determinate leading to the over-clad approach for the project, see Fig. 2. Erecting scaffolding to provide access to the interior would have added significant time and construction logistics problems to the project. This was successfully avoided by using the construction methodology of an over-clad installation for the project.

The outer skylight framing members are the primary support mechanism for the interior suspended lay-light system and all of the ceiling lighting systems. A conventional "T"-bar support for the lay-light system is suspended from heavy gage wire. Cylindrical "down-light" flood lamps and tubular fluorescent lighting was also supported by the lay-light grid, see Fig. 3. Leaving these systems intact throughout the construction of the replacement skylight led to the development of the "piggy-back" over-cladding approach used in the project.

System Description—The original skylight framing system consisted of extruded aluminum "I" sections with the top stem of the "I" comprising the bottom half of the glazing system, and included a receptor

FIG. 3—*A photograph of the interior between the inner lay-light and the outer skylight framing. Note that the outer skylight members are the primary support for the suspended lay-light structure and all of the ceiling lighting systems. Leaving these systems intact during construction was a primary consideration in the development of the "piggy-back" over-cladding approach used for the project.*

channel for the attachment of the exterior applied glazing cap. The bottom of the stem of the "I" included small vertical returns to create condensation control channels. These channels proved insufficient to adequately control drainage of condensate water from the noninsulated glazing. The primary "I" members were supported on continuous concrete curbs at the edge of the elaborate concrete roof forms. The framing elements formed complex curves in both plan and vertical cross section. A secondary interior ceiling framing system was suspended from the outer skylight framing system. The suspended system supported a series of interior lay-lights of obscure translucent glazing. Between the outer skylight and the suspended ceiling system, isolated down lighting for the sanctuary and fluorescent tubes were installed, which provided artificial "day-lighting." The result was a striking architectural aesthetic of both interior and exterior, both day and night.

The suspended ceiling system is located on a sloping plane from 30 to 55 feet above the floor of the sanctuary, and removal or reinstallation of the lay-light system from the interior would have been prohibitively costly, due to the extensive effort in providing access. Removal of the original outer skylight framing system would remove the support of the ceiling and lighting systems. Using an over-cladding approach to install a new skylight enclosure helped to avoid these significant construction and budget challenges.

Structural Issues—Changes in our understanding of wind behavior on structures since the 1960s have led to substantially higher code mandated wind loadings than were required in the original construction, and the new insulated glass units more than double the self-weight loads of the new skylight. A structural design check of the framing members of the original skylight system proved them inadequate for the support of the new system, increased code mandated wind and snow loads. A finite elements analysis of the concrete supporting structure proved that more than adequate reserve capacity existed for the new skylight system and its imposed loads.

The Solution—The retrofit skylight system specified required the following:

- A new state-of-the-art skylight glazing system with a proven track record of condensation and leakage control.
- Aluminum components of the new glazing system to be installed immediately above and work in tandem with the existing framing members to withstand the new gravity and wind loadings.
- Install the system from the exterior without requiring access to the full interior ceiling system, thus leaving the existing ceiling lay-light system as suspended from the original skylight framing members.
- Interior aluminum skylight components of the newly installed composite system to be thermally isolated from the exterior environment.
- New translucent laminated heat-strengthened glazing in lieu of the original wire-glass.
- Faithfully maintain the original expression of the significant architectural monument.

Mock-Up and Testing—Upon selection of the successful skylight contractor, in order to validate the proposed solution in the field, one of the 14 sections of skylight was selected as a full field mock-up. The primary glazing components of the over-clad system were standard proprietary profiles, developed by the system manufacturer, that had been prequalified for the project by prior testing and certification by standardized testing for air infiltration and water penetration in accordance with ASTM E 283-91 [1] and ASTM E 331-92 [2]. The structural portions of the extruded profiles were adapted for the over-clad application. After initial shop drawings and dies were approved and structural calculations verified the manufacturer's design, the primary components were extruded and erection access platforms installed on the building to perform a complete installation of one skylight section. Following initial mock-up installation, simple field testing was performed as an initial workmanship check. Initial prototype installation was completed in the late summer to allow more rigorous "testing" by Mother Nature in the form of fall rains and winter temperatures to validate the performance under real-world conditions. The intent was to begin the balance of the installation the following spring. By mid-winter, the performance was sufficiently validated in-place by significant thermal swings, heavy wind-driven rains, and normal midwestern winter snows. Production fabrication of the remaining 13 elements proceeded in late winter.

The original cross mullion system was considered state-of-the-art at the time of original construction. But it was glazed with uninsulated wire-glass; had limited water control mechanisms, and was prone to

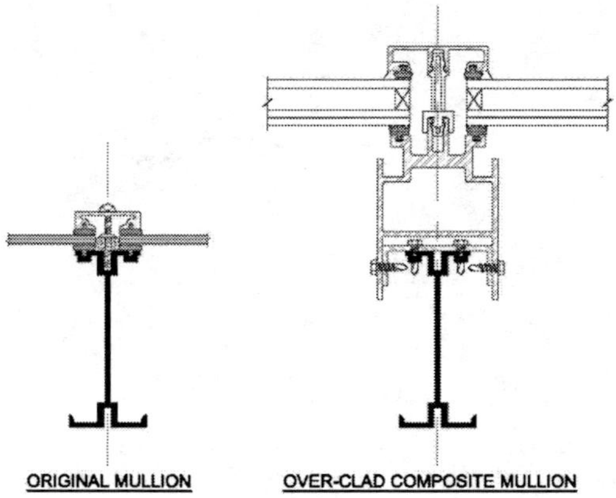

FIG. 4—*Old and new mullion assemblies. These figures illustrate the cross section of both the original system on the left, and the new over-clad skylight system on the right. Note the variations in the glazing sealant systems and condensation control systems (drainage channels). The lower half of the custom mullion section, below the outer "plumbing" components, was adapted specifically for the project to mate to the pre-existing "I" mullion below. These new components work in tandem with the original "I" section to provide the structural capacity necessary for the support of the modified system. The resultant three-component built-up mullion provided a mechanism to allow minor field adjustments to accommodate site tolerances.*

significant thermal (i.e., condensation) problems, subjecting the building to frequent bouts with water dripping to the spaces below, see Fig. 4. The modified system utilized more contemporary skylight glazing components for the support of new insulated glass elements comprised of heat-strengthened glass with laminated interior light. The interior aluminum components placed atop the original "I" mullion include proprietary glazing channels and condensation control mechanisms unique to the specific skylight manufacturer. This glazing system had a history of demonstrated performance for thermal (condensation) control, air infiltration, and water penetration, as demonstrated by physical testing to ASTM and related industry standards. The lower half of the custom mullion section, below the outer "plumbing" components, was adapted specifically for the project to mate to the pre-existing "I" mullion. These new components work in tandem with the original "I" section to provide the structural capacity of the new composite system. The three-component mullion built-up mullion allowed for minor field adjustments to accommodate site tolerances.

In order to adequately validate the construction methodology, verify field tolerance issues and site logistics, an initial field mock-up of one of the total 14 units was constructed ahead of the remaining skylight sections. Due to the nature of the cast-in-place concrete structure, a unique system of construction platforms was required to provide access to the skylight. This is illustrated in Fig. 5. Considerable measures were required to provide reasonable water-tightness of the structure during the installation of the over-cladding.

The interior lay-lights and ceiling lighting systems remained undisturbed, suspended from the original mullions sections. An interior view of the completed installation is illustrated in Fig. 6.

Upon completion, the original mock-up was subjected to simple water leakage testing, conforming to AAMA 501.2-83, which is currently being evaluated by ASTM Committee E6.55.15. The system was then left exposed to the elements for the balance of the fall rainy season, heavy winter snowfalls, and seasonal swings of temperature and humidity, as illustrated in Fig. 7. The resulting in-service exposure to the elements provided sufficient evidence to validate the installation.

Figures 8 and 9 contrast the appearance of the wall systems before and after construction.

Construction Phase—In the spring of the following year, production installation for the balance of the 14 skylight units commenced requiring additional installation platforms and production installation crews. The original site mock-up and field verification of dimensions allowed for shop fabrication of the remain-

FIG. 5—*A view of the initial field mock-up of one of the total 14 units that was constructed ahead of the remaining skylight sections is shown. Note the unique system of construction platforms required to provide construction access to the skylight, and temporary measures to provide water-tightness of the building during the course of construction.*

FIG. 6—*A view of the interior of the initial skylight mock-up section after erection, showing the new "piggy-back" mullion system. Note that the interior lay-lights and ceiling lighting systems remained undisturbed suspended from the original mullion sections throughout the construction phase.*

FIG. 7—*The original mock-up section of the project after completion. Following limited on-site performance testing, the system was left exposed to the elements during the fall rainy season and experienced heavy winter snowfalls and seasonal swings of temperature and humidity. The resulting in-service exposure to the elements provided sufficient evidence to validate the installation. Note the barely perceptible (nominally 4-in.) raised elevation of the completed skylight construction after modification.*

FIG. 8—*The general appearance of a typical segment of the original construction.*

ing skylight framing components and glass elements. A limited number of framing elements required site fitting to accommodate field tolerances and minor variations from segment to segment; a limited number of glass units required field measurement prior to fabrication. The skylight production installation was completed during a normal 40-hour work week, with the sanctuary available for services at weeks' end.

Comparison of the construction before and after the installation of the skylight retrofit illustrates the aesthetically sensitive manner in which the architectural expression of the building was maintained, while providing a wholly new functionality to the skylight, see Figs. 10 and 11.

FIG. 9—*A typical view of the replacement construction from a similar vantage point as Fig. 8.*

FIG. 10—*The original construction prior to construction as viewed from the south.*

FIG. 11—*A view of the construction from the same vantage as Fig. 10 following completion of the over-cladding installation.*

Case Study 1: Conclusion—The resulting skylight over-cladding solution proved to be the most economical and least disruptive solution to a very unique set of construction challenges. The newly installed skylight system remains a faithful expression of the original design architect's aesthetic vision, executed with contemporary skylight glazing components not available at the time of original design and construction. The installation provided a leak-free glazing system, wholly new thermal separation, "U" value and condensation control, increased wind and snow load capacity, and a more uniform visual appearance.

Case Study 2: Curtainwall Over-Clad

The second example of this "piggy-back" over-cladding approach is an upgrade of a late 1960s office tower with an original aluminum and glass curtainwall. The building was a fully occupied banking facility with single-pane glazing, and no thermal isolation of the metal components from exterior to interior. The design and construction challenge was to (1) correct structural deficiencies in the mullion anchorage system, (2) provide new energy efficiency to the wall system, (3) improve air and water penetration controls, while (4) providing a totally updated architectural expression for the building exterior, and (5) complete construction operations with minimal disruption to the existing interior office spaces.

System Description—The building structural frame is reinforced concrete. The original cladding system was an aluminum and glass curtainwall with exterior exposed aluminum "I" mullions, opaque glass spandrel panels, and tinted vision glass. The original mullions were two-story elements, gravity supported at mid-height. Inadequate provision was made in the original construction to accommodate normal vertical expansion/contraction of the mullion system created by temperature changes in the exposed aluminum mullions. Fortunately, copies of the original curtainwall vendor shop drawings were available for use in development of the new curtainwall system.

The owner felt that the building had a dated appearance (see Fig. 12), and desired a new expression for the building while upgrading the wall system to provide more contemporary performance than the original construction.

The original exterior mullion glazing system and glass prior to the recladding project had received numerous applications of remedial sealant between the glass and metal components, attempting to resolve ongoing water leakage of the exterior wall system, see Fig. 13.

In the original wall system configuration, the mullions were predominantly outboard of the glass line and exposed to the elements, See Fig. 14. The system had noninsulated glass and poor thermal properties and created cool areas adjacent to the exterior wall, which at times caused significant condensation at the interior.

Structural Issues—Site assessment of the building included removal of representative glass spandrel panels to allow observation of interior components of the wall system. These observations disclosed that portions of the original anchorage system were failing, due primarily to thermal movements of the wall system, and inadequate provision in the original anchorage system to accommodate these movements. These failures had apparently theretofore gone undiscovered. Analysis of the original gravity and wind load connections proved that they would be inadequate for the additional dead load of the insulated glass

FIG. 12—*The general appearance of the original building and curtainwall construction. The inset photo illustrates the appearance of the original exterior mullion elements and tinted vision and opaque spandrel glass.*

and the higher code defined wind loads of the proposed new curtainwall system. The mullion sections did have sufficient capacity to act as the primary gravity and lateral load carrying elements of the proposed over-clad curtainwall system. Analysis of the primary building floor framing proved to have sufficient capacity and stiffness for the additional loads imposed by the new curtainwall design.

FIG. 13—*An example of the condition of the original exterior mullion and glass systems prior to the recladding project. The presence of significant prior sealant application between the glass and metal components was indicative of many largely unsuccessful attempts at mitigating water leakage of the wall system.*

FIG. 14—*The appearance of the original wall system as viewed from the interior is illustrated. Note how the mullions are outboard of the glass line, exposing them to repeated thermal expansion/contraction, and causing significant thermal bridging to the interior.*

The Solution—The retrofit over-clad curtainwall solution required the following:

- Removal of the existing spandrel glass to allow installation of new primary structural anchorages for the pre-existing mullions to accommodate gravity and wind loads; original anchorages were then released.
- Leave the vision glass in place to enclose the structure while the new wall was erected to provide the primary weather enclosure of the building throughout the cladding erection phase.
- Installation of new custom fabricated glazing receptors exterior to the existing "I" mullions to provide the air and water control systems.
- Installation of new two-sided structurally glazed vision and spandrel glass units.
- Installation of new thermal installation, firesafing, and air control systems in the spandrel areas.
- Installation of the entire system without direct exposure of the interior environment.
- Upon completion of the exterior, remove all original vision glass from the interior, and complete interior trim installation in occupied spaces during off hours.

Fabricator/Erector Input—The project specifications were performance based and competitively bid. The wall system fabricator and erector awarded the contract had substantial input into the ultimate selection of the specific methodology for erection. This in turn had direct input into the specific component design, which was largely the result of the specific site erection logistics. Upon arriving at a system meeting all of the apparent needs and objectives of the project, fabrication drawings for a mock-up were prepared.

Mock-Up—Limited portions of the curtainwall, near the roof level, required extension of the original "I" mullion system to maintain the structural armature for the system. The fabricator chose to custom extrude mullion sections that exactly duplicated the original profiles for these supplemental elements. In doing so, the test mock-up truly representative of the eventual over-clad system for the project was able to be created at the fabrication site. ASTM testing of the assembly was performed at the fabricator's facility using ASTM E 283-99 [1], ASTM E 330-97e [4], and ASTM E 331-00 [5] to address air infiltration, water penetration, and structural capacity.

Construction—Upon successful completion of the specified preconstruction testing protocol, and shop fabrication of components, the contractor assembled standard modular scaffolding and erected the new cladding components during the fall and winter months, and began interior trim out in the spring. The system was erected from the top of the building downward to provide complete enclosure of the building, thus eliminating the considerable cost and inevitable site constraints that would otherwise have been

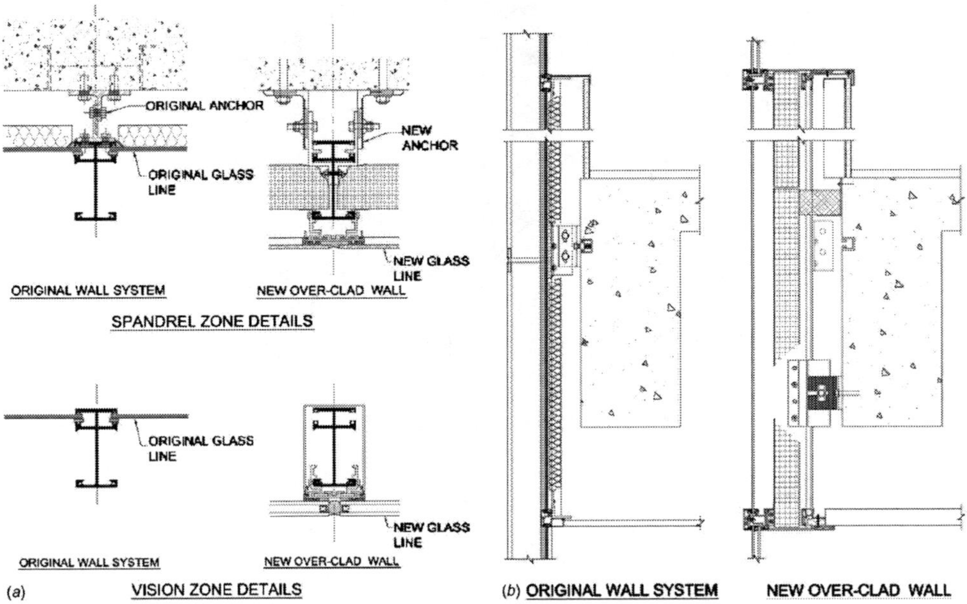

FIG. 15—*The wall system in plan view in its original configuration (left) and with the over-cladding elements in place (right). The upper half of the graphic shows the condition of the wall in the spandrel zones; the lower half, the vision glass areas. Note that in the modified wall system, the glass line has been moved to the exterior side of the original mullions sections, thus placing the mullion in-board of the glass in the retrofit wall condition.*

required for temporary weather enclosures and provision of supplemental heating.

Figure 15 illustrates the wall system in plan view in both the original configuration and with the over-cladding elements in place, at both the vision glass and spandrel zones.

Figure 16 illustrates how the new composite wall system moves the glass line to the exterior and

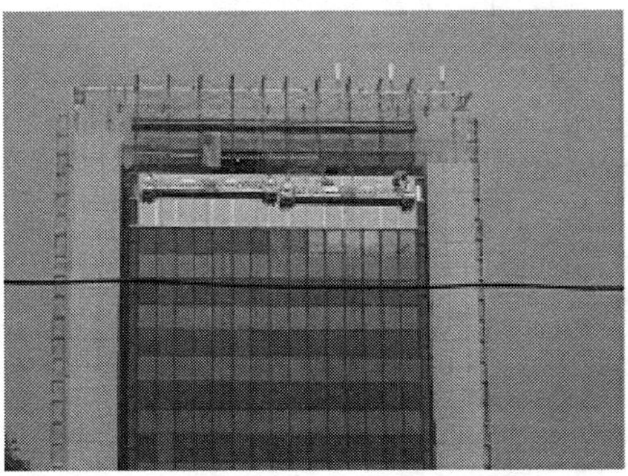

FIG. 16—*A vertical cross section of the wall in both the original (left) and modified condition (right). Note again how the new composite wall system moves the glass line to the exterior and provides significantly greater thermal, moisture, and air control mechanisms, as well as providing a very different architectural appearance.*

FIG. 17—*How the new over-clad system was installed from the top down, allowing continuous temperature and weather control of the interior environment is shown. Note how the vision glass was left in place as the over-clad system was installed, thus allowing erection to proceed with minimal displacement of the tenants during construction. The photo illustrates the significant change in architectural expression of the building exterior wall after erection of the new components. The transformation from old to new is illustrated, showing the uncovered wall, a partial completed section, and fully completed section at the top.*

provides significantly greater thermal, moisture, and air control mechanisms, as well as proving a wholly new architectural expression.

Erection Sequence—The erection operations proceeded in the following sequence:

1. Original spandrel glass and spandrel insulation was removed from the exterior to expose the anchorage zone.
2. New structural anchors were installed beneath the original anchors; the original anchors were then released.
3. New glazing profiles were attached to the outer flange of the original exposed structural "I" mullions.
4. Spandrel zones were insulated.
5. New spandrel and vision glass was installed.
6. Exterior structural glazing completed and exposed caps and architectural fins installed.
7. Original vision glass was removed from the interior side and interior trims installed.

The new over-clad system was installed from the top down to allowing continuous temperature and weather control of the interior environment, see Fig. 17. The vision glass was left in place while the over-clad system was installed, allowing the erection to proceed with minimal displacement of the tenants during construction. The new glazing system provides a completely new architectural expression of the building exterior.

The modified wall system moves the glass to the exterior face of the mullions, and creates a modified appearance of the completed wall, both exterior and interior. After erection of the new glass and glazing components at the exterior, the original vision glass was removed from the interior; interior mullion covers, head, and sill trims were installed. In the finished installation, none of the aluminum components of the original wall system remain visible, see Fig. 18, yet they still provide the primary load resisting capabilities of the new retrofit system.

The new wall system included the addition of applied vertical architectural fins at the exterior on the completed wall to create a vertical architectural expression to the building, see Fig. 19. The resulting retrofit construction provides a striking contrast to the original wall system, and fully disguises the fact that the primary aluminum mullion components of the original wall remain under the new skin of the building.

FIG. 18—*A view of the newly installed wall components viewed from the interior. Note how the glass line has moved outboard of the mullion. The interior mullion covers, head and sill trims are all new, and in the finished installation, none of the aluminum components of the original wall system remain visible.*

FIG. 19—*The new architectural expression of the wall system after erection of the new components. The new construction provides a striking contrast to the original wall system, while disguising the fact that the primary aluminum mullion components of the original wall still remain as a significant part of the new exterior wall of the building.*

Case Study 2: Conclusion—The curtainwall over-cladding design conceived on paper, translated near perfectly onto the existing exterior wall system in the field, and the resulting construction is a fully functional, modern, and energy efficient curtainwall system, that was erected under the curious eyes of the building occupants, while causing only minimal disruption of the building interior spaces. The new wall system has the appearance of a fully new wall system, both interior and exterior, belying the fact that the bulk of the structural capacity of the new wall system is provided by the original curtainwall components.

Over-Cladding as Environmentally Friendly Construction

Over-cladding solutions, such as the two illustrated here, are environmentally friendly solutions...

- By reusing primary elements of a pre-existing system in tandem with new components; less new raw materials are required for the new installation. This translates into considerable energy saving through economizing on the use of new raw materials, and the necessary energy to translate those new materials into finished forms.
- By recycling existing wall components in-place in their original form and extending their useful service life, substantial energy is saved that would be required to put otherwise discarded materials back into use through conventional recycling, (i.e., recycling without additional energy consumption).
- In situations where a new installation can be installed without exposing the interior building environment to the exterior elements, significant temporary heating or cooling costs can be eliminated, resulting in substantial construction phase energy savings, as well as eliminating direct costs of providing temporary enclosures.
- By turning thermally inefficient, single-glazed curtainwall systems into energy conserving systems through the introduction of insulated glass, thermal insulation, energy saving glass coating systems, and introduction of new insulating materials, over-clad systems can have significant effect on long-term energy consumption of the building.

Future Opportunities

The two examples shown illustrate a basic concept of bringing poor performing and inefficient cladding systems up to date with comparable new construction technologies, while addressing real-life construction logistics inherent in fully occupied buildings. These concepts can be easily translated to a large variety of mid to late 20th century buildings clad in a variety of materials. As energy sources become more scarce and related costs continue to escalate, competition in the marketplace drives owners of our existing building inventory to compete against newer more up-to-date construction, recladding of buildings will likely become more commonplace. Over-cladding is one important option available to accomplish this important part of building modernization.

Acknowledgments

The author wishes to acknowledge the efforts of the Naturalite Skylight Division of the Vistawall Group, X-Clad, Inc., Chicago Height Glass, Berglund Construction Company, the office of Minoru Yamasaki, Wright Architects, and Mr. Richard Fencl for their cooperation and their respective roles in the development of the design solutions executed on the projects represented in the case studies. It is through their close collaboration that these projects were successfully executed.

References

[1] ASTM Standard E 283, "Standard Test Method for Rate of Air Leakage of Exterior Windows, Curtain Walls and Doors by Uniform Static Pressure Difference," *Annual Book of ASTM Standards*, ASTM International, West Conshohocken, PA, 1991.

[2] ASTM Standard E 331, "Standard Test Method for Water Penetration of Exterior Windows, Curtainwalls and Doors by Uniform Static Pressure Difference," *Annual Book of ASTM Standards*, ASTM International, West Conshohocken, PA, 1993.

[3] ASTM Standard E 283, "Standard Test Method for Rate of Air Leakage of Exterior Windows, Curtain Walls and Doors by Uniform Static Pressure Difference," *Annual Book of ASTM Standards*, ASTM International, West Conshohocken, PA, 1999.

[4] ASTM Standard E 330, "Standard Test Method for Structural Performance of Exterior Windows, Curtain Walls and Doors by Uniform Static Pressure Difference," *Annual Book of ASTM Standards*, ASTM International, West Conshohocken, PA, 1997.

[5] ASTM Standard E 331, "Standard Test Method for Water Penetration of Exterior Windows, Curtainwalls and Doors by Uniform Static Pressure Difference," *Annual Book of ASTM Standards*, ASTM International, West Conshohocken, PA, 2000.

Journal of ASTM International, Vol. 4, No. 9
Paper ID JAI100870
Available online at www.astm.org

Shawn S. Li[1] *and Neil McClelland*[2]

Façade Renovation of the Wexner Center for the Arts

ABSTRACT: The Wexner Center for the Arts went through a major façade renovation to fix the water leakage and condensation problems. The work included replacing glazed skylights, curtain walls, and roofs. This paper describes how renovation work has been done with modern technologies to study thermal and lighting performance on the new façade. In addition to the computational analysis, a full scale mock-up was constructed in a laboratory for air, water, and structural test, and to evaluate the condensation resistance performance. Currently there are no industry standards for condensation evaluation for a mock-up of this size. The mock-up test method refers to ASTM C 1199, "Standard Test Method for Measuring the Steady-State Thermal Transmittance of Fenestration Systems Using Hot Box Methods," and AAMA 1053 test procedures but with modifications to suit the project. The new façade of the Wexner Center is a perfect example of how façade design can be done with modern analyses tools and a coordinated design approach to deliver an environmentally efficient cladding system. The paper also describes the issues for the current test standards to the projects that have specific design conditions.

KEYWORDS: Façade, Condensation, THERM, Thermal Test

Introduction

The Wexner Center for the Arts was built in 1989. Designed by architect Peter Eisenman, the building soon became an artistic statement and a landmark building of The Ohio State University. The building won a National Honor Award in 1993 from The American Institute of Architects. Its exterior consists of glass curtain walls on the east, south and north elevations and a skylight over the main gallery. The outside of the eastern façade features distinctive steel scaffoldings that connect back to the building's main structure by penetrating the curtain wall and skylight (Fig. 1).

Because of its innovative architectural design and the large use of glass to the building envelope, the façade technologies and related computer programs available at the time could not provide a sustainable solution and could not predict the performance of the building façade. Just a few years after the completion, the Wexner Center experienced many problems including condensation and environmental control. The glazing units let in too much daylight, which could potentially damage art works. The skylight also began to leak. Unable to fix the problems, the Wexner Center eventually covered the skylight with membrane, and applied translucent plastic film to the curtain wall glazing. These temporary measures turned a dazzling gallery into a gloomy, unpresentable space.

FIG. 1—*The Wexner Center for the Arts.*

Manuscript received October 23, 2006; accepted for publication August 23, 2007; published online September 2007.
[1] Senior Façade Engineer, PE, LEED AP, Arup, New York 10013
[2] Principal, Arup, New York 10013

The gallery space is required to maintain the interior environmental conditions at 50 % relative humidity and 70 °F with very small fluctuation at any time to preserve the long-term life of the artwork. In the winter, the high relative humidity caused major condensation problems on the curtain wall. The local contractor's attempts to fix the problem could not stop the reoccurrence of the condensation.

Eventually, the Wexner Center administration decided to undertake a major renovation to remediate the problems with long-term solutions. The initial proposal was to redesign the mechanical system, but after much review, it was decided that a greater scale of renovation was required. The final scope of the renovation, relevant to façade issues, included HVAC, curtain wall, skylight, and roofing.

Façade Thermal Performance

Condensation has often been overlooked in past building designs. The damage done by condensation does not often appear instantly, but instead usually occurs over time.

However, condensation itself does not necessarily equal problems. The formation of condensation has several stages starting from a layer of mist, to small droplets and eventually forming large droplets. A layer of mist that dissipates in a few hours should be acceptable to most buildings.

It is safe to assume that condensation is likely to occur on a surface if the surface temperature is below the dew point temperature. ASHRAE (American Society of Heating, Refrigerating and Air-Conditioning Engineers) has recommended keeping the surface temperature of the glass and frames a minimum 5 °F above the room dew point for a high humidity environment [1].

While changing the interior environmental condition is not an option in this project because of the museum environments, there are only two ways to keep the façade surface temperature above the design dew point. One is to provide a localized heating source, such as running hot water conduit or electrical heat trace through the metal frame. The other is to reduce the heat conduction between the inside and outside by introducing a thermal barrier between the interior and exterior. The former method can become problematic because of the potential for water leakage and localized over heating. The latter option is a more effective method and has been widely used in today's façade technology.

A few methods can be used to determine the thermal performance of a façade system. Computer simulation and mock-up testing are the most common ones. Among many computer simulation programs, THERM is the most popular in the United States. Developed by Lawrence Berkeley National Laboratories (LBNL) and approved by the National Fenestration Rating Council (NFRC) [2], THERM is based on the finite element method and can simulate two-dimensional steady state heat-transfer effects in fenestration systems and other building envelope constructions such as walls, foundations, roofs, and doors. THERM is used in conjunction with programs Window 5 and Optics 5 (also developed by LBNL) to model the thermal performance of façade components.

For the mock-up test, ASTM, AAMA, and NFRC have guidelines for fenestration thermal testing, which uses thermal couples to measure the temperature difference between weather side and room side on an approximately 80 in. by 80 in. test specimen [3–6]. AAMA and NFRC also provide procedures for determining the Condensation Resistance Factor (CRF) [4,7]. These standards are useful for glass and frame system selection and comparisons, but are not suitable for projects with specific project designs and environmental conditions.

Façade Design for the Wexner Center

Computer Simulation

Climatic data that was used for the exterior design conditions for the Wexner Center was based on ASHRAE climatic design information [8]. The information provided by ASHRAE is intended for the mechanical system design while also serving as a good reference for a fenestration condensation analysis. The ASHRAE design heating temperature for 99.6 % dry-bulb is 1 °F. This means that the exterior design conditions for the Wexner Center are likely to have a temperature equal or lower than 1 °F for 35 hours each year. After reviewing the local historical weather data and considering the sensitivity of the gallery spaces to the effects of condensation, −5 °F was chosen by the design team for the winter external design temperature.

TABLE 1—*THERM boundary conditions comparisons.*

	Current Standards		Wexner Simulations	
	Exterior	Interior	Exterior	Interior
Ambient Temperature (°F)	0	70	−5	70
Radiation Temperature (°F)	0	70	−5	65
Surface Air Film (Btu/h-ft^2-°F)	5.28	1.35	4.58	0.55
Relative Humidity, Dew Point	N/A	N/A	N/A	50 %, 50.5°F

THERM comes with the predefined boundary conditions, which is the same as the specified testing environmental conditions of AAMA 1503 and NFRC 102 (Table 1). The predefined conditions are convenient for thermal assessments and comparisons for glazing products. The program also allows users to create new customized boundary conditions. However, this has to be done with caution. When THERM is used to simulate the glazing system for a building, the inputs for boundary conditions such as ambient temperature, surface air film coefficient, and radiation temperature will need to be adjusted to suit the building's façade and HVAC design.

Because the Wexner Center's curtain wall and skylight are adjacent to each other, the glazing surface can "see" itself. For this type of façade, the boundaries are more complex and the designed interior environmental conditions cannot be simply applied to THERM computer simulations [9,10]. To determine more realistic boundary conditions for THERM models, the project team used Computational Fluid Dynamics (CFD) to evaluate air flow, façade surface ambient temperature, and radiation temperature throughout the entire building (Fig. 2). The CFD results were then applied to THERM models for façade thermal analyses.

Table 1 shows the comparison of the boundary conditions between the current testing standards (AAMA 1503 and NFRC 102) and the Wexner Center:

Façade System Selection

The original skylight and curtain wall system of the Wexner Center consists of a nonthermally broken frame and insulated double glass unit. Both glazing and frame are largely under designed, given the high humidity environment.

The requirements for the new façade system were to have an adequate thermal performance throughout the glazing and frame system while still maintaining the original look. Potential curtain wall systems that met these requirements were built into THERM models and simulated for comparison (Fig. 3).

The simulations suggested that a regular double glazing unit with a selected thermal broken frame, under the design conditions, would have an average performance but the risk of condensation was still

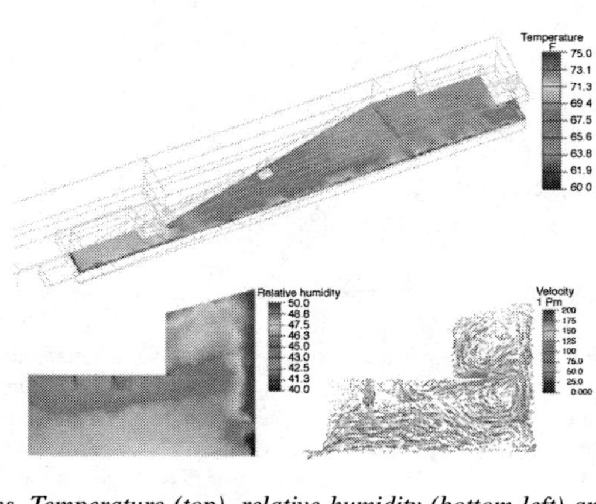

FIG. 2—*CFD simulations. Temperature (top), relative humidity (bottom left) and air flow (bottom right).*

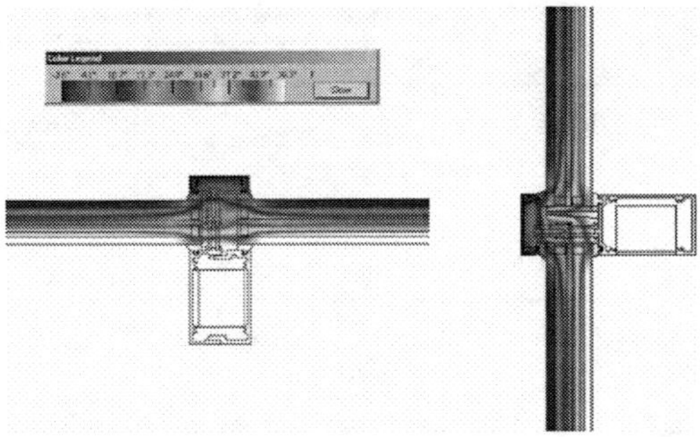

FIG. 3—*THERM simulation results; jamb (left) and sill (right).*

high. The computer analysis led to a triple insulated glazing unit, and the final framing system consisted of a gasket isolator and strip-and-clamp thermal break of the two-level thermal barrier system that has a CRF rating between 74 to 80.

The glazing selection created another challenge for the Wexner Center. In order to meet both the thermal and daylight requirements, the glazing units were required to admit a proper amount of visual light, limit solar heat gain for thermal comfort, and block ultraviolet light to protect the artwork. Day lighting studies were carried out to locate the "hot-spots" from direct sunlight and to establish the interior illuminance level from the diffuse light.

The selection process went through numerous thermal and daylight analyses to determine the best combination of aesthetics, lighting, and thermal performance (Fig. 4). The typical project glazing assembly included a triple glazed unit with laminated glass in a clear or translucent interlayer. In the end the project had a total of 13 different types of glazing assemblies. The make-up of the typical triple pane unit is shown in the Table 2.

Table 3 is the approximate u-value for the façade system.

FIG. 4—*Daylight factor analyses.*

TABLE 2—*Glass make-up for Wexner Center.*

Outer Pane	1/8-in. heat strengthened with reflective coating on inner surface + PVB interlayer + 1/8-in. heat strengthened
Air Gap	1/2-in. argon fill with stainless steel spacer
Middle Pane	1/8-in. heat strengthened with low-e coating on inner surface
Air Gap	1/2-in. argon fill with stainless steel spacer
Inner Pane	1/8-in. heat strengthened + PVB Interlayer + 1/8-in. heat strengthened

TABLE 3—*U-value for the typical façade system.*

Components		u-value (btu/h-ft^2-°F)
Curtain Wall	Frame	0.38
	Center of Glass	0.17
	Overall	0.25
Skylight	Frame	0.5
	Center of Glass	0.23
	Overall	0.3

Mock-up Test

The preparation of the façade construction work started in the fall of 2004. At the same time, full-scale mock-up testing was carried out at a testing laboratory. The test specimen included a typical curtain wall elevation connecting to a skylight. In addition to the water, air, and structural test, a thermal test was also performed on the same mock-up. As explained in the previous section, the condensation resistance factor (CRF) is useful for product selection and comparisons but not for performance evaluation in a specific environment. Therefore, the main purpose of this thermal test was not to obtain the thermal transmittance or the CRF, but to examine the formation of condensation on the new façade system. ASTM C 1199, AAMA 1503, and NFRC 102 have developed testing procedures to evaluate the thermal transmittance of fenestration systems [3,4,6]. AAMA 1503 and NFRC 500 have the procedures of determining the CRF [4,7]. All these standards are similar to each other. The approximate 80 in. by 80 in. glazing wall test specimen was placed in a test chamber. The test boundaries were set at a predefined environmental condition as defined in Table 1. While the standard procedures are good for measuring the thermal transmittance for typical fenestration systems, they are not appropriate for the determination of the actual occurrence of condensation [7], and cannot accommodate the specific conditions of the project. The CRF has also tented to underestimate the requirements [10,11].

The best way to access the condensation risk is a performance test that is similar to the air, water, and structural tests. Currently, there are no such standard guidelines for testing a full-scale customized glazed wall.

The project team referred to the standards mentioned above and established the testing procedures. The testing environmental condition was set at the design condition, which is 70°F, 50 % RH for the inside, and −5°F for the outside. Air flow was created to achieve the required surface air film coefficient, but the speed was not strictly monitored because it is difficult to control a steady air flow for a mock-up of this size.

Following the air, water, and structural tests, the mock-up was enclosed with insulation boards (Fig. 5), and 56 thermal couples were placed at critical locations for comparing with the calculated dew point. The test chamber was stabilized to the required conditions and maintained for at least two hours. The temperature and humidity levels were recorded in three-minute intervals.

During the test, a thin layer of condensation formation occurred on the edge of the glass. While there are not industry standards to define the pass and fail of the test, the thin layer of mist was considered acceptable as it did not form into droplets. Condensation water was also observed on the underside of the head member at the intersection of the wall and skylight. Further inspection found a metal angle that was

FIG. 5—*Full-scale thermal mock-up test.*

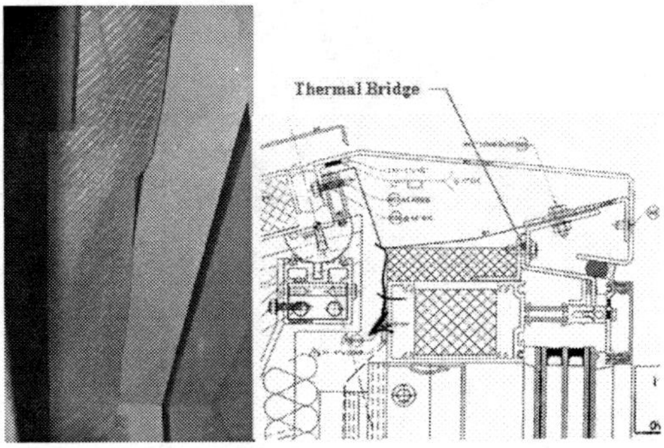

FIG. 6—*Condensation from the test and the thermal bridge detail.*

used to support the skylight structure and created a thermal bridge (Fig. 6). A new detail was then proposed by the engineering team to eliminate the cold bridge. Further THERM analyses were carried out to confirm the remedial method (Fig. 7).

Overall, except for a few irregularly low measurements, the interior temperatures recorded from the mock-up test were 2 to 9°F different from the interior temperature from THERM simulations.

Conclusion

The new façade of Wexner Center is an example of modern façade design. It is an effective integration of architecture, structure, mechanics, and lighting. The new curtain wall and skylight extend the exhibition experience to a new level and bring the Wexner Center back to its original purpose (Fig. 8).

Condensation causes not only aesthetic issues but also health concerns from the induced mold growth and potential structural deterioration. Condensation issues have normally not played a significant role in the selection for curtain wall systems. When it comes to condensation control, façade designers and engineers often rely on CRF or simply address the issue by relocating the vapor barrier. Detailed condensation analysis is generally not considered in the design process. Façade engineers tend to concentrate on structure and details for water leakage control, while mechanical engineers usually focus only on façade design to meet building energy code requirements. With more condensation problems reported, condensation analysis is now recognized by more designers and engineers. This is especially true for buildings that require special interior environments such as museums, hospitals, and natatoriums.

CRF alone will not tell us what stage or what amount of condensation will form on the surface of the façade at specific design conditions. Computer simulations can be good references for initial evaluation,

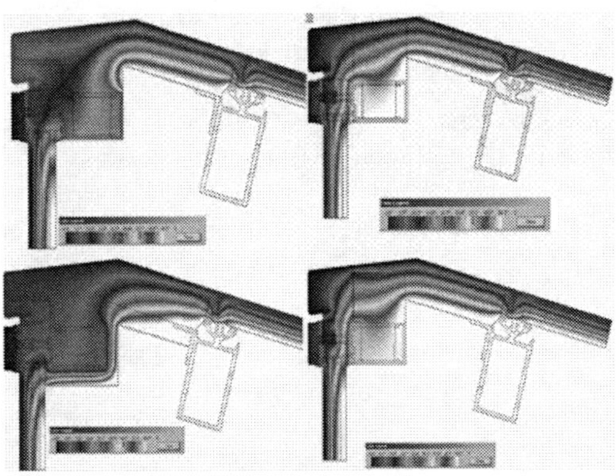

FIG. 7—*THERM simulations on different detail options.*

FIG. 8—*The Wexner Center façade before (left) and after (right) renovation.*

but mock-up testing will be of great help for more accurate determination of the condensation formation.

Except for the condensation resistance factor from AAMA 1503, there are no other current standards or guidelines for condensation evaluation. Establishing a standard testing procedure for condensation risk assessment is in an urgent need among the industry.

References

[1] ASHRAE, *HVAC Applications Handbook*, Chapter 4: Places of Assembly, 2003.
[2] NFRC 101, "Procedure for Determining Thermo-Physical Properties of Materials for Use in NFRC-Approved Software Programs," 2004.
[3] ASTM Standard C 1199-00, "Standard Test Method for Measuring the Steady-State Thermal Transmittance of Fenestration Systems Using Hot Box Methods," *Annual Book of ASTM Standards*, ASTM International, West Conshohocken, PA.
[4] AAMA 1503.1-98, "Voluntary Test Method for Thermal Transmittance and Condensation Resistance of Windows, Doors, and Glazed Wall Sections."
[5] NFRC 100-2004, "Procedure for Determining Fenestration Product U-Factors."
[6] NFRC 102-2004, "Procedure for Measuring the Steady-State Thermal Transmittance of Fenestration Systems."
[7] NFRC 500-2004, "Procedure for Determining Fenestration Product Condensation Resistance Values."
[8] *ASHRAE Handbook Fundamentals*, Chapter 26: Climatic Design Information, 1997.
[9] Curcija, D., "Analyzing Thermal Performance of Building Envelope Components Using 2-D Heat Transfer Tool with Detailed Radiation Modeling," CEERE.
[10] Kohler, C., "THERM Simulations of Window Indoor Surface Temperatures for Predicting Condensation," *ASHRAE Transactions*, Vol. 109, Pt. 1, 2003.
[11] O'Brien, S. M., "Finding a Better Measure of Fenestration Performance: An Analysis of AAMA Condensation Resistance Factor," *Building Envelope Forum*, 6th Issue, August 2006.

Journal of ASTM International, Vol. 5, No. 7
Paper ID JAI100865
Available online at www.astm.org

Tammy D. Forner[1]

A Form to Follow Function: Recladding A Regional Healthcare Facility

ABSTRACT: This case study presents the recladding program undertaken by Riverside Regional Medical Center in Newport News, VA, and examines the deliberative process that resulted in retrofit of an existing 40+ year old aluminum/glass/metal panel curtain wall system without removal of the existing system or opening large areas of the façade to exterior weather. Steps within the process included: (1) establishing a design approach and air/water control theory; (2) verifying structural capacity of the existing façade anchorage and vertical structural members; (3) identifying and specifying products that met the design and performance requirements; (4) evaluating an innovative option for improved thermal performance; (5) testing the air/water barrier to verify performance; and (6) careful monitoring of work progress and installation quality. All of this was done with very close coordination with hospital management and staff to minimize service disruptions and occupant discomfort for this sensitive client population. The final result includes increased energy efficiency and acoustical performance of the façade system—while presenting a progressive new image to the public.

KEYWORDS: building façade, curtain wall, recladding, retrofit, air/water barrier, rain screen cladding

Introduction

The Riverside Regional Medical Center is a 576-bed hospital in Newport News, Virginia. The original hospital building was built in 1961, with a second building added in 1965. The two buildings vary in height up to seven floors, with a roof level penthouse on one building and a connector link between the two buildings. The façade for both buildings featured two cladding systems: (1) brick masonry on concrete block backup at building end walls, one-story podium structures, First Floor columns between storefront openings, and secondary façades at the rear of the hospital; and (2) an aluminum/glass/insulated steel "sandwich" panel curtain wall system at main facades from the Second Floor through the Roof. While there were no reported building façade air/water control concerns, exposed exterior aluminum curtain wall trim had become dull and pitted over time. The insulated steel panels appeared to have been repainted at various times during the life of the building and, in 2001, were beige in color. Overall, the building appeared drab and dated (Figs. 1 and 2). Additionally, the hospital is located within 10 miles of Langley Air Force Base and the Naval Base at Norfolk, VA, and is subject to a high amount of military jet traffic and the associated noise.

The hospital engaged a design and construction team in the fall of 2001 with a goal of improving façade performance and updating the appearance of the buildings. Initial discussions were centered on removing/replacing cladding on the entire building façade, including complete replacement of the curtain wall system. The discussion soon turned to overcladding the entire existing façade, and replacing single-glazed windows with thermally efficient insulating glass windows. The existing façade would be left in place behind the new façade.

As the design evolved, a glass-and-metal-panel façade became the favored esthetic option. Re-use of the existing curtain wall as a platform for support of the new cladding at curtain wall areas was advanced as the desirable structural support solution, allowing most of the work on the building to proceed from the building exterior and eliminate opening the existing façade to weather elements during installation. Thus, structural adequacy of existing curtain wall system and its anchorage to meet updated building code windload requirements became the central focus of study.

Manuscript received May 12, 2007; accepted for publication May 15, 2008; published online June 2008.
[1] Senior Design Consultant, Curtain Wall Design and Consulting, Inc. (CDC, Inc.), New England Office, 64 Maple Avenue, Suite 1, Canton, CT 06019.

FIG. 1—*Riverside Hospital, September 2001: partial West elevation, original hospital building.*

FIG. 2—*Riverside Hospital, September 2001: partial East elevation, original hospital building.*

Existing Curtain Wall and Structural Systems

The existing curtain wall system was non-thermally broken, extruded aluminum vertical and horizontal framing, with $\frac{1}{4}$ in. (6 mm) clear glass used at vision areas and insulated metal sandwich panels at spandrel areas. Architectural and Structural drawings of the original building documented a steel framed building with steel deck-and-concrete composite floor slabs. Curtain wall shop drawings from hospital archives showed curtain wall anchorage affixed to the structural steel floor beams near edge of slab at the Third Floor through Roof, and an independent structural channel to receive curtain wall starter anchors installed at the Second Floor. Curtain wall mullions spanned two floors; therefore, windload/deadload anchorage was made at the Second Floor and Fourth Floor, with windload-only anchors located at the Third and Fifth Floors and the Roof.

Initial Investigation and Analysis

In August 2001, concurrent with design option discussions, an initial investigation was undertaken to verify that the existing anchorage was consistent with that shown on the shop drawings. Further, condition

TABLE 1—*Windload design criteria, per ASCE 7.*

Code Considerations	Condition Definition
Building height	75 ft (22.86 m)
Wind exposure	Exposure C
Building openings condition	Condition I (Enclosed); no impact resistant glazing required in the lower 60 ft (18.29 m)
Importance factor	III (Emergency Facility)
Wind speed	85 mph (136.79 kph)
Exposure area	50 ft^2 (4.65 m^2)
location	30 miles (50 k) from hurricane coastline
Parapet condition	3 ft (0.91 m) parapet around the roof perimeter

of the curtain wall anchorage was observed. These initial observations confirmed that the size of anchor clips, number of anchor bolts, and location of typical anchors on the building were consistent with the shop drawings, and that the anchors showed no deterioration that was of structural concern.

During September 2001, the Newport News Building Department was consulted to ascertain accepted criteria for establishing windload for the façade system under BOCA code, which was the applicable code at that time in Newport News. Based on these discussions, criteria listed in Table 1 were used to determine wind design loads. Wind loads were determined to be +39 psf/−41 psf (+1.9 kPa/−2.0 kPa) at typical façade areas, and +39 psf/−49 psf (+1.9 kPa/−2.4 kPa) at corner zones.

Structural analysis of existing mullions as detailed in archival shop drawings verified that they were sufficient to resist the new windloads without deflecting beyond the industry standard L/175. Structural analysis of the anchors shown on the shop drawings determined that they complied with the new windload criteria. Assuming no structural deterioration of anchors, analysis determined that no structural enhancement for resistance to wind pressures would be required at anchors in typical façade areas; and at corner façade zones, only the windload/deadload anchors at the Fourth Floor would require enhancement.

A concurrent examination of adjacent brick cavity wall construction was undertaken to determine what kind of anchorage might be required if these areas were overclad as well. Due to budget considerations, this part of the work was later abandoned in favor of maintaining the masonry façade and replacing only the windows within the masonry.

On-Site Testing to Confirm Structural Adequacy

The addition of the overcladding elements added weight to the existing curtain wall, and moved the center of gravity outbound from that of the existing curtain wall system. The existing curtain wall anchorage to the building structure was load tested to verify anchor and mullion structural capacity to resist windload and confirm that anchors would resist additional deadload and eccentricity. The condition of anchor assemblies at a representative number of locations was also verified during testing. Table 2 lists the

TABLE 2—*Results of anchor testing.*

Floor #	Room #	Windload R_{xn}, lb (kg)	Deadload[a] R_{xn}, lbs. (kg)	Δ_{max} Anchor, in. (mm)	Δ_{set} Anchor, in. (mm)	Δ_{max} Mullion, in. (mm)	Δ_{set} Mullion, in. (mm)
2	245	600 (272.2)	-0-	0.010 (0.254)	0.000	0.010 (0.254)	0.000
2	225	2000 (907.2)[b]	-0-	- -[b]	- -[b]	- -[b]	- -[b]
3	325	2200 (997.9)	-0-	0.010 (0.254)	0.000	0.090 (2.286)	0.060 (0.524)
3	351	2200 (997.9)	-0-	0.170 (4.32)	0.025 (0.635)	0.100 (2.54)	0.090
4	445	...	...	...	...	...	...
4	Annex	1300 (589.7)	350 (158.8)	0.020 (0.508)	0.000	0.020 (0.508)	0.000
4	425	1300 (589.7)	350 (158.8)	0.005 (0.127)	0.000	0.060 (1.524)	0.015 (0.381)
4	ICU	1300 (589.7)	350 (158.8)	0.000	0.000	0.080 (2.032)	0.010 (0.254)
5	545	2200 (997.9)	-0-	0.010 (0.254)	0.000	0.020 (0.508)	0.005 (0.127)
6	ICU	2200 (997.9)	-0-	0.000	0.000	0.000	0.000

[a]Applied at 3-5/8 in. (92 mm) outbound of centerline of mullion anchorage.

[b]Load gauge malfunction; no observed damage.

FIG. 3—*Jack stand, load cell, and attachment bracket used in testing anchors; angle extension from bottom of attachment bracket provided location for application of 350 lb (158.76 kg) dead load.*

calculated wind reaction loads, based on worst-case tributary areas at each type of anchor on the building, and the results of anchor testing. The testing process is summarized as follows.

Metal spandrel panels at typical wall areas and corner zones were removed to expose anchors for observation and testing. Observations revealed that the anchors were in good condition. Several locations exhibited some rust; however, the amount of rust observed did not represent a compromise to structural capacity.

A custom jack stand and an attachment fixture were employed to simulate wind reaction loads at various floor levels within the typical and corner anchorage zones. The test procedure included applying loads to anchors, holding for 10 s, and recording deflections at mullions and steel brackets relative to the floor beam; then the load was released and any residual deflection noted. A Com-Ten[2] Series 301 Roof Fastener Tester with 4000 lb (1814.4 kg) capacity was employed to apply loads. Deflections were measured with Ames[3] dial indicators calibrated to 0.001 in. (0.025 mm).

An additional test was applied at the Fourth Floor windload/deadload anchors, to simulate additional deadload and eccentricity of loading due to the overclad system. Testing conservatively assumed that the new metal panel and window systems would add 5 lb/ft^2 (0.76 kg/m^2) weight to the building, with center of gravity at 3-5/8 in. (92 mm) outbound of the centerline of mullion anchorage, and additional deadload of 350 lb (158.76 kg) at each anchor. During testing, deadload was applied simultaneously with the simulated windload. (Fig. 3).

Finalizing Esthetic and Performance Criteria

Architectural documents issued in January 2002 reflected an operable replacement window at glazed areas, with a rain screen metal panel system and rigid insulation board applied to the face of the original metal panel portions of the facade. Original façade materials and the selected recladding/replacement materials are presented in Table 3.

The new aluminum panels would be installed on new aluminum channel supports applied to the face of existing curtain wall mullions. The new panel system featured two panel depths that projected from the face of existing mullions 5-3/4 in. (146 mm) at the main face of panel system and 8-3/4 in. (222 mm) at accent panels. New, thermally efficient pre-glazed operable windows were to be adapted to insert into existing glass-holding members.

[2]Com-Ten Industries, 6506 49th St. N., Pinellas Park, FL 33781.
[3]B. C. Ames, Inc., 78 Stone Place. Melrose, MA 02176.

TABLE 3—*Performance comparison, original façade materials and recladding materials.*

	Original Façade Materials		Recladding/Replacement Materials	
Material	Manufacturer and Product	Technical Performance Data	Manufacturer and Product	Technical Performance Data
Metal panel areas	No data available; 3 in. (75 mm) thick metal sandwich panels with steel outer skin, filled with rigid fiber board insulation	$R=12$[a] (RSI=2.11)	Pohl Europanel and Ecopanel products; $\frac{1}{8}$ in. (3 mm) thick aluminum sheet[b]	$R=n/a$. Rain screen panels offered no insulating value
Rigid polystyrene insulation	N/A; not part of original construction	N/A	Dow Corning CavityMate™ Styrofoam™ insulation[c]	3 in. (75 mm)thick, $R=15$ (RSI=2.65)
Windows	Integral to aluminum framing system, non-operable	No data available; note that aluminum framing was not thermally broken or thermally improved	Wausau 3250Z[d]	Condensation Resistance Factor (CRF)=44[e] U-Value: 0.69 BTU/ft2/h/°F (3.91 W/m2/°K)[e]
Glazing	$\frac{1}{4}$ in. (6 mm) clear monolithic glass	Solar Transmittance: 77 % U-Value: 1.09 BTU/ft^2/h/°F (6.17 W/m^2/°K) Winter; 1.03 BTU/ft^2/h/°F (5.84 W/m^2/°K) Summer Shading Coefficient: 0.94 Solar Heat Gain Coefficient: 0.81 STC: 31[g]	1 in. (24 mm) insulating glass units consisting of $\frac{1}{4}$ in. (6 mm) PPG Azurlite® with Solarcool® (2)[f] – $\frac{1}{2}$ in. (12 mm) airspace – $\frac{1}{4}$ in. (6 mm) clear glass	Solar Transmittance: 12 % U-Value: 0.48 BTU/ft^2/h/°F (2.72 W/m^2/°K) Winter; 0.50 BTU/ft^2/h/°F (2.83 W/m^2/°K) Summer Shading Coefficient: 0.29 Solar Heat Gain Coefficient: 0.25 STC: 35[g]

[a]Value for 3 in. thick rigid fiber board insulation available in current market.
[b]Christian Pohl GmbH, Robert bosch STR. 6, D-50769, Cologne, Germany.
[c]Dow Chemical Co., 2030 Dow Center, Midland, MI 48674.
[d]Wausau Wall and Window Systems, 1415 West Street, Wausau, WI 54401.
[e]When tested per AAMA Voluntary Test Method for Thermal Transmittance and Condensation Resistance of Windows, Doors and Glazed Wall Sections (1503–98).
[f]PPG Industries, One PPG Place, Pittsburgh, Pennsylvania 15272.
[g]Commonly accepted value for 1/4 in. (6mm) monolithic and 1 in. (25 mm) insulating glass units with no laminated glass included in the assembly.

An Innovative Approach to Air, Water, and Thermal Performance

An innovative alternative approach to adding thermal insulation also enhanced the air and water barrier plane of the façade. SuperTherm®[4] as manufactured by Superior Products International II, a product line developed for the coatings and roofing industry, was recommended as a substitution for rigid insulation board. When cured, the product system formed a flexible, UV-, air-, and water-resistant coating.

Application for this project included use of a reinforced mastic tape membrane to encase existing vertical and horizontal curtain wall framing members at their joints with the metal spandrel panels. Tape membrane was also used to seal the replacement windows to the existing curtain wall system. The membrane was coated with a fluid-applied, protective base coat, followed by a water-resistant insulating coating, roller-applied over the membrane as well as the face of existing spandrel panels. The insulating coating material contained ceramic materials suspended in a urethane and acrylic base that blocked both infrared and ultraviolet transfer of heat. Published insulation values of R-19 (RSI-3.34) at 8 mils dry film thickness, exceeded the performance of the originally specified 3 in. (75 mm) insulation board while providing better continuity of installation than rigid insulation could achieve.

[4]Superior Products International II, Inc., 10835 W. 78th St., Shawnee, KS 66214.

TABLE 4—*Insulation substitution technical review concerns.*

Insulation Installation Concern	Manufacturer Response
Membrane susceptibility to UV exposure	Insulating coating must be applied within 30 days of membrane application
Installation temperatures - membrane	Manufacturer recommended installation when air temperatures above 70°F (21°C); prescribed special application procedures for temperatures <70°F (21°C)
Assured adhesion of membrane	Field observation/testing led to use of roller tools to insure membrane adhesion
Acceptable substrate condition	Clear definition of the term "clean" established the manufacturer, and recommended substrate cleaning methods prescribed
Sequence of application	Defined to insure no discontinuities (Fig. 4) or reverse shingling that might encourage entry of water behind products
Dry film thickness of insulating coating	DeFelsko PosiTector 6000 Series Model FNS3[a] coating thickness gauge used for verification; installer reporting forms illustrated average coating thickness as specific locations
Accessory product compatibility	Via manufacturer testing, Pecora 895[b] identified for use in sealing insulation system terminations, penetrations and discontinuities.

[a]DeFelsko Corporation, 802 Proctor Avenue, Ogdensburg, NY 13669.
[b]Pecora Corporation, 165 Wambold Road, Harleysville, PA 19438.

The product substitution was rigorously examined through the submittal process, prior to acceptance. Critical issues reviewed are summarized in Table 4. Once all questions were answered, this alternative was ultimately accepted by the construction and design team, and construction commenced in Spring of 2003.

Performance Testing

A test mock-up area was erected on the hospital building in May 2003. Preparation for test mock-up included installation of all elements required to establish air- and water-barrier continuity, omitting installation of the rain screen metal panels. The area prepared as test specimen was two floors in height and four windows in width.

Initial testing was performed at a group of three adjacent windows within the prepared area, in accordance with AAMA "Field Check of Water and Air Leakage Through Installed Exterior Windows, Curtain Walls and Doors by Uniform Air Pressure Difference" (501.3-94) modified to exclude testing for air infiltration rate; and ASTM Standard E1105 "Field Determination of Water Penetration of Installed Exterior Windows, Curtain Walls, and Doors by Uniform or Cyclic Static Air Pressure Difference" [1]using uniform static air pressure difference of 10 psf (0.479 kPa). The test resulted in no observed infiltration of uncontrolled water behind the air/water barrier plane, and was declared a success. A repeat of this test at a different location on the building was performed on Aug. 27, 2003. (Fig. 4). Testing was again run in accordance with AAMA 501.3-94 and ASTM E1105, at 10 psf (0.479 kPa) uniform static air pressure differential, and again resulted in no observed infiltration of uncontrolled water behind the air/water barrier plane.

Monitoring of Work Progress and Installation Quality

Field quality control was a chief concern of the hospital and the construction manager/general contractor. A rigorous field monitoring system was enacted by the subcontractor, including item-by-item checklists for installation of the replacement windows, mastic tape membrane/fluid-applied air barrier and insulation, and metal panel system. The subcontractor employed three full-time quality control inspectors to monitor all aspects of the installation on a day-to-day basis, and to keep records of their observations. Third-party site observation visits were made on two- and four-week cycles to verify that work was proceeding in

FIG. 4—*Test chamber installed at West elevation of building, for second test of air/water barrier, Aug. 27, 2003. Note "shingling" of dark-colored mastic tape membrane installed to the left of the test chamber.*

accordance with the parameters set during shop drawing/tech data review and initial performance testing, and to review field quality control reporting with the construction manager/general contractor and the subcontractor.

Items monitored during field installation were:

1. Sequence of membrane installation to insure a "shingled" installation (Fig. 4).
2. Identifying/sealing discontinuities after application of fluid-applied insulating material.
3. Cleaning of metal surfaces after removal of existing operable windows (Fig. 5).
4. Membrane and sealant installation during and after installation of replacement windows (Figs. 6–10).
5. Thickness of fluid-applied insulation material at 8 mm minimum dry film thickness.
6. Full sealing of panel system anchorage fasteners through the air/water barrier plane (Fig. 11).
7. Size and placement of fasteners from metal panel carrier clips to vertical panel carriers (Fig. 12).
8. Expansion joints between vertical panel carriers.
9. Metal panel engagement onto vertical panel carriers.

FIG. 5—*Condition of window opening after removal of existing window. Note remains of butyl-based sealant, glazing tape residue, and dirt. Cleaning process involved use of organic solvent cleaning agents and vigorous scraping with putty knives to remove old sealant, glazing tape, and dirt.*

FIG. 6—*Interior window trim with pre-applied mastic tape membrane is applied to cleaned window opening.*

10. Sealant joints at system terminations (Second Floor starter sill and Roof Coping).

Overcladding at Connector Link

The connector link between the Main Hospital building and the addition required a different approach to façade update. The original cladding was a corrugated industrial sheet metal system mounted on a steel strap backup that was attached back to structural steel via welded steel clips, and contained narrow vertical

FIG. 7—*Horizontal pressure bars replaced, and new vertical pressure bars applied. Opening is ready to receive additional layer of mastic tape membrane and new pre-glazed window.*

FIG. 8—*Window attached via screws at new vertical pressure bars.*

FIG. 9—*Additional mastic tape membrane applied at perimeter of window unit, adhered starting from bottom and working upward in a "shingled" application.*

FIG. 10—*Completed mastic tape membrane application; location is ready for application of primer and insulating coating.*

FIG. 11—*Metal panel carrier clip with fasteners penetrating air/water barrier; all clips set in bed of compatible sealant and fastener heads oversealed.*

FIG. 12—*Metal panel system carrier clips, awaiting installation of panel carrier channels. All fasteners from clips to carrier channels had to be verified for type and placement, and provisions for carrier channel expansion verified prior to installation of panels.*

FIG. 13—*Base connection of corrugated metal panel system and vertical window strips at connector link. Note extensive water damage and deterioration.*

FIG. 14—*Connector link during overcladding with glass and aluminum curtain wall. Note anchorage access openings in metal panels following slope of structural beams beyond.*

FIG. 15—*Riverside Regional Medical Center today.*

strips of windows inserted at intervals along the façade. Due to poor sealing of both windows and metal panels, and inadequate maintenance, the system had been subject to water infiltration for many years.

During visual observation, it was apparent that the existing system would not support the panel system applied at the other two buildings. Water infiltration had deteriorated the metal panels and their base connections. Structural support of these panels was deemed insufficient to support loads imposed by a curtain wall system under current code requirements (Fig. 13).

The decision was made to leave the metal panel system in place; open panels at locations that would allow access for anchorage to structural steel beyond; and overclad the metal panel system with a standard aluminum-and-glass curtain wall system (Fig. 14). This allowed the connector to remain operational and relatively weather tight during installation of the glazed system. When completed the existing metal panel system was completely encased by the glazed system, protected from further deterioration due to water infiltration, and freed from structural loading due to wind.

Conclusion

At project completion in late 2004, the result proved to be worthy of the effort. The likelihood of maintenance issues stemming from water infiltration through the façade is reduced via the metal panels preventing most (if not all) rain from ever reaching the air/water barrier plane. Installation of windows with energy-efficient tinted low-E insulating glass has upgraded both thermal and acoustical performance of glazed areas. Overcladding the existing metal panel façade at the connector "link" has eliminated concerns over water infiltration as well as structural integrity of that system. A drive past the facility reveals a progressive and modern façade, in keeping with the image of a modern health care facility (Fig. 15).

References

[1] ASTM Standard E1105, "Field Determination of Water Penetration of Installed Exterior Windows, Curtain Walls, and Doors by Uniform or Cyclic Static Air Pressure Difference," *Annual Book of ASTM Standards*, Vol. 04.11, ASTM International, West Conshohocken, PA, 2004.

Journal of ASTM International, Vol. 5, No. 8
Paper ID JAI100876
Available online at www.astm.org

C. Hendryx[1] *and S. Vlotho*[1]

Repair of an Earthquake Damaged Building Façade in Coordination with FEMA: A Case Study Including Stabilization, Evaluation, Historically Sensitive Restoration, and Hazard Mitigation

ABSTRACT: The Nisqually Earthquake struck Washington State in February 2001, damaging the facade of the historic Bush Hotel, located in Seattle's International District. During the seismic event, sections of exterior cement plaster parge cladding and clay block infill fell from the structure onto the sidewalk below. This paper outlines the failures of the 1920s era concrete frame and hollow-core clay block infill exterior wall assembly, the multi-faceted approach designed to implement a historically sensitive repair in coordination with FEMA regulations, and addresses mitigation of potential future hazards. Prompt evaluation of the life safety threat was conducted, and sprayed-in-place polyurethane foam was used as a unique method to stabilize and temporarily weatherize portions of exterior walls. Various techniques were used to evaluate exterior walls and quantify the extent of damage including hammer sounding, invasive testing, mapping, and CAD-assisted measurements in a rigorous process to meet FEMA funding requirements. Damage to the exterior walls included delaminated and cracked stucco cladding, crushed hollow-core clay block infill panels, and cracked and spalled concrete. A multitude of factors were addressed in the Construction Documents including: restoring exterior walls to pre-earthquake conditions; maintaining the historical fabric of the building's appearance; coordination with FEMA and historic review agencies; repair to hollow-core clay block without in-kind material readily available; restoration of an exterior mural; performing work in a busy downtown setting with the building fully occupied; and addressing removal of lead-containing coatings. Importantly, a plan for mitigation of potential hazards was designed and implemented utilizing anchors to secure clay block infill panels to the structure, in order for wall sections to better resist possible future earthquakes. Methodology and examples for the synthesis of well-coordinated Contract Documents for such a complicated project are addressed. Evaluation of successes and challenges in coordination with FEMA, design, and repair methods are also provided.

KEYWORDS: Public Assistance Program, Emergency Assistance Act, earthquake repair project, hazard mitigation, masonry repair, concrete repair, hollow-core clay block, hollow clay tile, cement plaster (i.e., stucco) repair, elastomeric coating, mural restoration

Background

On February 28, 2001, Western Washington experienced a seismic event that damaged the exterior walls of the Bush Hotel, located in the International District of Seattle, Washington. Seattle Chinatown International District Preservation and Development Authority (SCIDPDA), the nonprofit organization managing the low-income apartment building, retained the authors' firm to assist in evaluation of the damaged exterior walls, develop a scope of repair including drawings, specifications, and related contract documents, while administering the project through the protocols and procedures required by the Federal Emergency Management Agency (FEMA).

Constructed as a single room occupancy residential project, the Bush Hotel primarily housed Seattle's Asian immigrant population. Today, the building provides low-income housing. Constructed with a six-story cast-in-place reinforced concrete skeleton frame structure and in-filled with hollow-core clay block, the exterior was covered in a three-coat cement plaster parge. The hollow-core clay blocks, set in a lime-based mortar, served as the substrate for both the exterior and interior finishes. There is no weather barrier within the exterior wall assembly; instead the nonreinforced infill panels, which are one to three

Manuscript received October 21, 2006; accepted for publication July 8, 2008; published online August 2008.
[1] Technical Manager and Project Manager respectively, Building Envelope Technology and Research, 4000 Delridge Way SW, First Floor, Seattle, WA 98106.

FIG. 1—*Undated period photograph of the Bush Hotel.*

wythes thick, manage water as a mass wall. Overhanging metal cornices partially protect the primary walls from the elements.

Prior to the earthquake in 2001, the building owners had completed a five million dollar renovation of the facility during the mid-1990s. The renovation was executed in accordance to the guidelines of the National Park Service Historical Renovation Standards, as the Bush Hotel is listed in the national historic registry and is situated in a historic district of the city. Figure 1 illustrates an undated historic photograph of the Bush Hotel. During the 1990s extensive renovation, the exterior cladding was spot-repaired and repainted, roof perimeter parapet walls were seismically braced, and shear walls were constructed in locations (on the interior side of select exterior walls) required for safe egress, using shotcrete to reinforce the infill panels. Importantly, it was later established during the Earthquake Repair Project that the earlier renovation in the 1990s provided the baseline for the "pre-disaster condition" of the building.

After the earthquake, the owners of the Bush Hotel needed help to evaluate the damaged facility, establish eligibility under FEMA public assistance regulations, and administer a repair project subject to significant bureaucratic requirements. During the Earthquake Repair Project, the owners and their project team worked to balance many factors including project budget and financing constraints, governmental policy, and the technical challenges of the repair work. Figure 2 further illustrates various considerations encountered during the project. In order to strike this balance, a thorough understanding of the regulatory requirements for public assistance funding was important, and thus is outlined in the following section.

Survey of the Public Assistance Program, and Coordination with Federal Emergency Management Agency (FEMA)

The Federal Public Assistance Program was established to provide help for the repair of buildings and other facilities damaged during disasters. Following a declared disaster, such as an earthquake, an owner of a damaged building may be eligible for public assistance for repair costs. In order to apply for FEMA relief, the owner must first determine whether they are an eligible entity and whether the building is an eligible facility according to FEMA regulations. If determined to be eligible, and therefore able to apply for public assistance, there is then significant work required to identify the scope of repairs that may be covered and then administer the relatively complicated processes. The following outlines aspects of the Public Assistance Program, especially as they pertain to the Bush Hotel Earthquake Repair Project. The outline here is not all inclusive, and the authors strongly suggest that the owner and design team evaluate

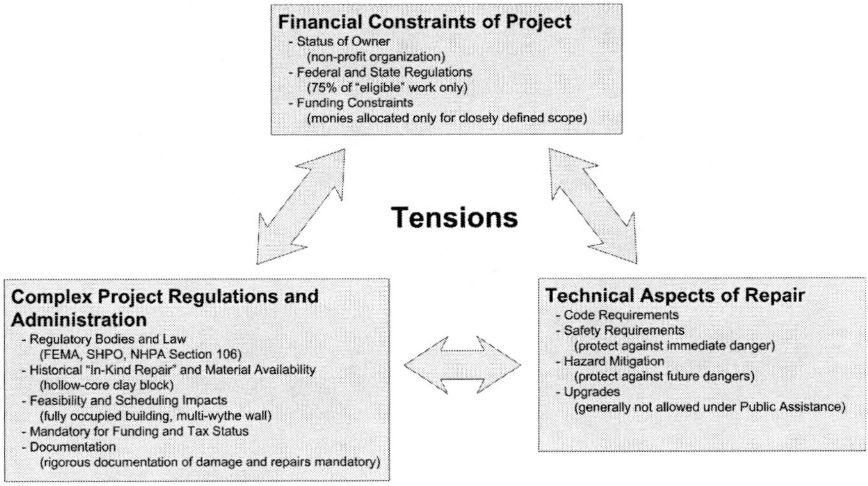

FIG. 2—*Diagram depicts the tensions between the competing considerations in a repair project funded through the Federal Public Assistance Program.*

FEMA and State publications[2] and contact their State Public Assistance Officer as soon after a disaster as possible. Regulations are complex, and therefore understanding key documents, regulations, and establishing a relationship with your State Public Assistance Program representative are important first steps.

Federal Statute and Public Assistance Program

The Bush Hotel Repair Project was funded according to The Robert T. Stafford Disaster Relief and Emergency Assistance Act (Stafford Act), 42 U.S.C., Public Law 93-288. It is through the Stafford Act that the Department of Homeland Security's Federal Emergency Management Agency (FEMA) provides funding for emergency work to State agencies [1], and it is the statute that authorizes Federal Public Assistance for the repair or replacement of facilities damaged as a result of earthquakes and other declared disasters. Furthermore, regulations published in Title 44 of the Code of Federal Regulations (44 CFR) Part 206 implement and interpret the Stafford Act statute. The regulations outline the program procedures, as well as rules for eligibility and funding [2]. Additionally, policies issued by FEMA clarify aspects of the regulations, provide information for specific situations, as well as explain various intricacies of the Public Assistance (PA) Program [3].

In order to fund a project, a Federal grant is awarded by FEMA to the "Grantee," the State government. The Grantee is accountable for the use of the funds provided by FEMA and is responsible for distributing those funds to the Sub-grantee, the Applicant. (The actual State agency that fills the role of Grantee, and as the prime administrator of the PA Program, varies from state to state.) For example, in Washington State, the Grantee is the Washington State Military Department Emergency Management Division. The Grantee is also responsible for providing technical advice and assistance to eligible Applicants, ensuring that all potential Applicants are aware of the available assistance programs, providing support for damage assessment operations, as well as submitting the necessary paperwork for grant awards [4].

Figure 3 outlines the general relationship between the Federal, State, and local entities, and the organization that applies for public assistance, i.e., the Applicant:

[2]Publications pertinent to the Federal Public Assistance Program include: FEMA 321: *Public Assistance: Policy Digest*, Federal Emergency Management Agency, Washington, DC, 2001, FEMA 322: *Public Assistance: Public Assistance Guide*, Federal Emergency Management Agency, Washington, DC, 1999.
Refer to State-specific publications as well, for example: *Washington State Public Assistance Manual*, State of Washington Military Department Emergency Management Division, Olympia, Washington, 2006.

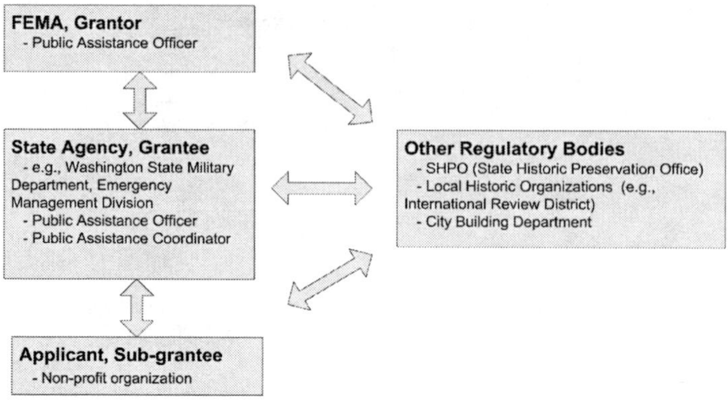

FIG. 3—*Diagram depicts general relationships of parties involved in Public Assistance Program.*

Work Eligibility Under Public Assistance Program

As stated, following a disaster, a building owner must first establish eligibility for the Public Assistance Program. It is important that early after the disaster, the eligible status of the owner's organization and the status of the damaged facility are established with the State. Under the Stafford Act there are a variety of entities eligible for Federal Public Assistance. For example, the Bush Hotel, which provides low-income housing, is operated by a private, not-for-profit organization. Private nonprofit organizations (PNPs) may be eligible for public assistance, based on several factors, and within limitations. For example, PNPs are eligible only after applying for a loan through the Small Business Administration, and collecting all applicable insurance monies [5].

The Federal Public Assistance Program organizes eligible work into two categories, "Emergency" and "Permanent," which are further subdivided into subcategories:

Emergency (Emergency protective measures)

"A" Debris clearance

"B" Protective measures

Permanent

"C" Road systems

"D" Water control facilities

"E" Building and equipment

"F" Public utility systems

"G" Parks, recreational facilities, other items

Eligible repair work includes work to restore a facility to only "pre-disaster" design, function, and capacity. Conversely, betterments of design (e.g., more conservative bracing), change in facility use (e.g., conversion of a meeting hall into offices), and change in facility capacity (e.g., increasing size and seating in meeting hall), are strictly not allowed, and furthermore, these "betterments" can jeopardize funding. During the Bush Hotel Project, the proposed repairs were carefully evaluated by FEMA and the State to ensure that betterments were not included.

Although betterments must be avoided, upgrades to the facility to meet current building codes are allowed within limits; however, they are carefully scrutinized. The Applicant must show that the upgrades:

- Apply only to the elements that were damaged by the disaster;
- The code upgrades were appropriate for the pre-disaster use of the facility;
- The code standard is reasonable, and was formally adopted and enforced prior to the disaster;
- The code standard is applied uniformly to all facilities of a similar type in the area [6].

As described later, improvements in the building's design and construction are allowed within explicit parameters in order to prevent certain future potential life safety hazards.

Applicant Process

Following a disaster, an owner must become aware of the deadlines that need to be met for the Public Assistance Program. Within 30 days of a disaster designated by the Federal Government, an Applicant must submit a Notice of Interest (or Request for Public Assistance) to the State agency. Specific project information outlining eligible repairs must then be submitted to the State and FEMA within 60 days of the initial "kick-off" meeting. FEMA is obligated to respond to the request for public assistance within 45 days, and then the Applicant is granted another 60 days to submit an appeal if necessary. An appeal may be necessary if:

- For example, State/Grantee approved costs are less than the Applicant believes is necessary to complete the work;
- Portion of cost claimed for the work is not eligible;
- Applicant disagrees with approved scope of work in the Project Worksheet (PW).

Establishing Eligible Repairs and Forecasted Cost of Project

If the Applicant, facility, and damage to the facility, are all approved by FEMA as eligible, FEMA then works to determine the eligible repairs and cost of the project, and therefore the amount of Federal funds to be obligated to the project. Currently, a project estimated to be over $50 600 qualifies as a "large project." Typically, as long as repair costs are less than 50 % of the value of the facility, repairs are mandated in lieu of replacement. Note that PNPs are only eligible for repair projects.

Policy indicates that large projects are ultimately funded based on actual costs of the eligible work—in other words, the cost of the project may be increased during the project by following strict guidelines.

An overview of the process of administrating a large project includes:

- The FEMA Project Officer prepares the Project Worksheet (PW), which describes the damage, scope of eligible repairs and related work, and estimates the total project cost including construction, engineering and design, and other costs;
- As repair work progresses, the Applicant requests progress payments from the State based upon documented, completed work. Note that this process adds a delay in the funding—an important point that cannot be overlooked when working to pay the repair contractor under the terms of the Contract Documents;
- At project completion, the State surveys the work, and performs inspections and audits, or both. The final cost of the project is determined and the State submits a final report to FEMA, documenting completion and requesting payment.

Following the disaster, the Applicant must clearly demonstrate the disaster-caused damage that requires repair. Documentation of the disaster-caused damage to the facility is important for proving that a project is eligible for a public assistance, as well as securing funding through the State and FEMA for repair of the damage. Upon first reading of public assistance informational publications, it may seem that defining and agreeing upon "damage" is straightforward. Representatives of FEMA will visit the facility, and make a preliminary determination of the extent and cause(s) of damage. However, depending on the size and complexity of the project and the nature of the damage, it is the authors' experience that the PNP must be prepared to engage a reputable consulting firm to provide substantial investigation and analysis of the damage, and provide detailed documentation. Without thorough documentation, the Applicant may not be able to sufficiently "substantiate" the damage requiring repair.

Often, the Applicant must bear most of the cost for documenting disaster-caused damage. "Building inspections" by the PNP and their consultants are only eligible for reimbursement if the inspections are conducted in order to establish whether the damaged structure poses an imminent life safety threat. Whereas typically, inspections to establish quantity and mode of damage are not eligible, and therefore a nonprofit agency cannot be compensated for what may possibly be a relatively expensive survey. However, as stated, a detailed survey is necessary to provide accurate documentation of damage, in order to procure FEMA funding. Inevitably, it may be determined that each square foot of a building cannot be surveyed and instead compromises may be made, e.g., survey of a representative sample of exterior of locations and extrapolation of the results. Without a thorough survey, the PNP is confined to rely on FEMA's sole opinion and judgment of the damage(s). With the results of a complete and independent survey however, a PNP can negotiate or challenge the type and extent of damage with FEMA as may be necessary, through the appeal process.

As previously stated, the public assistance policy states that for large projects the actual cost of repairs will be compensated, and thus, a detailed independent survey may not initially be perceived as necessary. However, detailed surveys can often more clearly establish the mode, type, and extent of disaster-related damage. More complete information regarding the damage ensures that the FEMA Project Worksheet is more comprehensive, thus helping to eliminate potential discrepancies and reasons for undue limitation of funding at a later date. In addition, more complete information about damage also helps the preparation of more accurate Contract Documents, helping to streamline contract administration during repairs.

After determination that a project is eligible, establishing eligible damage, and defining a repair scope of work, FEMA will require validation of cost estimates that are included in the Project Worksheet cost estimate. The cost estimates must be shown to be reasonable and necessary to accomplish the work, including historical cost documentation for similar work on other similar projects, average cost for the geographical area, published unit costs, and in addition, FEMA cost codes. This is especially important in locals that have higher than average construction costs. Furthermore, the scope of work must be shown to be in compliance with Federal, State, and local regulations. Documentation must be provided by the Applicant to show that the cost of the project has been reduced by all applicable credits, such as loans and insurance payments. Finally, importantly, the PNP organization must be prepared to fund 25 % of the project. Repair projects for PNPs are mandated at 75 %–25 % cost sharing between the Federal Government and the PNP organization.

Bidding and Contract Award

Following establishment of the estimated project cost in the PW, the project may then be designed, and issued for public bid. FEMA highly encourages contract procurement for large projects to be through sealed bids or competitive proposals, and it is mandatory that contract procurement follow all Federal requirements. Noncompetitive proposals and negotiated contracts are discouraged. Contract types can be lump sum, unit price, or cost plus fixed fee. The selection of a contract type generally is dependent on the information known about the damage and necessary repairs. A lump sum contract may be beneficial if damage is well defined, whereas if the quantity of damage, and thus quantity of work, is less well known, a unit price-based contract could prove useful.

Examples of Special Project Considerations

There are other special considerations made during planning of a public assistance repair project for a PNP. Two such considerations in the Bush Hotel project were "hazard mitigation," and restoration of an exterior wall mural.

Section 406 of the Stafford Act outlines hazard mitigation, defined as cost-effective action taken to prevent or reduce the threat of future damage to a facility. Upgrades required to meet current codes and standards are not considered hazard mitigation measures for purposes of the PA Program, and have different eligibility criteria. Hazard mitigation must be:

- Applied only to the damaged "element" of the building;
- For hazard mitigation measures to be approved, the measures must be reviewed by FEMA staff to ensure eligibility, technical feasibility, environmental and historical compliance, and cost effectiveness;
- Hazard mitigation measures may amount to up to 15 % of total eligible cost for the project;
- Mitigation measures may be determined to be cost effective as long as the mitigation measure does not exceed the cost of the eligible repair work on the project [7].

Federal funding may also be used to stabilize artwork affected by a disaster. The funding is applicable for a "minimum of steps necessary to return an item to condition it can function in the same capacity as it did prior to disaster." FEMA's Preservation Officer, in consultation with Applicant and State, will use professional judgment to determine if additional treatment beyond stabilization is necessary to maintain the integrity of an artwork, and return it to its pre-disaster function. If integral to a historical building, the historical significance of the artwork will be evaluated, and will be incorporated into the historical review and preservation effort.

Regulatory Requirements

FEMA must ensure that projects under the Public Assistance Program comply with Federal and State laws and regulations, for example, the National Environmental Policy Act (NEPA), and the National Historic Preservation Act (NHPA). For example, in compliance with Section 106 of NHPA, FEMA must:

- Identify historic properties;
- Evaluate the effects of the PA Program projects on historic properties;
- Consult with the State Historic Preservation Office (SHPO), the Advisory Council on Historic Preservation, and other interested parties;
- Obligate funding only after completing the historic review process [8].

Evaluation of the historical impacts of the intended repair project is coordinated with the State Historical Preservation Office, and any local entities with jurisdiction.

Project Management and Execution

A strong focus on project management, administration, and documentation is fundamental to ensure the requirements of the PA program are met, and that funding of the project is therefore secured. It is critical to have thorough documentation, for validation, approval, funding, and audit processes. As the repair work progresses, the project will be periodically inspected by FEMA and State representatives in order to verify that eligible repairs have been made, proper documentation of the repairs is taking place, and in order to authorize a progress payment for the work accomplished. Thorough, organized, and accurate records of repairs, work completed by the contractor, design and administration costs, helps to make the progress approvals expedient. The Applicant, repair contractor, and consultant must also be well prepared for an audit at the project's end—complete and organized documentation can reduce the potential for problems encountered during the audit process, whereas incomplete documentation can lead to the reduction or elimination of funding.

While proceeding with the project, the Applicant is responsible for ensuring that the grant funds are used only for the explicitly stated eligible work. Any changes to the scope of eligible work, cost of the work, or increase in the duration of time required to complete the work, must be immediately reported to the State, and then FEMA, before the change takes place. If the change is not authorized in writing, prior to nonapproved work taking place, or prior to the elapse of an established deadline, funding may be jeopardized. After reviewing the State's report regarding a proposed change, FEMA will consider adjusting the amount of the grant to reflect the actual cost of the eligible work. If approved, the Project Worksheet for the project is revised and reissued. This can occur several times during the course of a challenging project.

The requirement for reporting changes in the work is especially important for conditions that are considered "hidden damage." During the course of making repairs to eligible damage, new types of damage may be uncovered, or a greater extent of a known type of damage may be documented. The Applicant is required to show how the damage is related to the original disaster-event. Hidden damage is considered for eligibility on a case by case basis, and will be inspected by FEMA's representative with the Applicant. Authorization to proceed with the repair project could be delayed if additional review is required under NEPA or NHPA [9]. It is highly recommended that the Applicant, repair contractor, and consultant have a plan in place to accommodate the additional time required to administer the approval of potential hidden damage.

Assessment and Documentation of Damage

With a magnitude of 6.8, the forces of the 2001 Nisqually Earthquake were generally directed through the Seattle area in an east-west direction. During the earthquake, debris consisting of unbonded cement plaster (stucco) and damaged hollow-core clay block fell from the exterior walls of the Bush Hotel to the ground. SCIDPDA, the nonprofit organization managing the building, assembled a consulting team consisting of a structural engineer and envelope consultant to provide initial survey of the damage and specify temporary emergency repairs. Emergency removal of additional loose cement plaster and clay block from exterior walls was conducted immediately.

The cast-in-place concrete, skeleton frame structure experienced the greatest forces and deformation primarily at the north and south elevations. Seismic forces acting on the concrete frame resulted in temporary deformation of the structure during the earthquake, magnifying the stress at the corners of the hollow-core clay block infill walls at exterior walls. The relatively thin and brittle webs of the unreinforced hollow-core clay blocks were crushed, causing delamination of the cement plaster parge coat over clay blocks and over the concrete frame, posing a potential threat to the health, safety, and welfare of the building's occupants and general public. The publication FEMA 154, *Rapid Visual Screening of Buildings for Potential Seismic Hazards: A Handbook*, states, "Where URM [unreinforced masonry] infill is present, a falling hazard exists" [10]. (FEMA's definition of the concrete frame with unreinforced masonry is illustrated on page 27 of the FEMA 154 document [11]). In another document, FEMA lists the failure mechanisms typically associated with unreinforced masonry:

- Bed-joint sliding;
- Diagonal cracking;
- Corner compression;
- Out-of-plane failure [12].

All of these modes of damage were observed on the Bush Hotel project.

The original structural design had been engineered for the construction of additional floors, and the initial post-earthquake survey indicated that the reinforced-concrete structure encountered no major structural damage as a result from the seismic event. During the repair project, only limited damage to the concrete structure itself was discovered.

Following the earthquake, holes in the exterior wall assembly were visible at the north, east, and south elevations, and outward bulges of failed cement plaster parge coat were visible at all elevations. Open voids in the exterior wall existed where fractured hollow-core clay block and unbonded cement plaster cladding had fallen away. To provide further protection to pedestrians, the owners erected scaffolding protection over the street-level storefronts and sidewalks. And, in accordance with guidelines for emergency work eligible under the Public Assistance Program, open voids in the wall assembly at the north and south elevations were filled with commercial-grade, two-part spray polyurethane foam. The temporary emergency repair was intended to consolidate fractured tile together, restore weather resistance of the exterior wall, and help prevent further falling debris in the worst damaged areas. Open voids in the exterior wall were first cleaned of debris, masked off to protect adjacent cladding, filled with spray-applied foam, and then coated to protect the foam and match the color and texture of the existing cement plaster. The foam was planned to be removed during future, permanent repairs. Installation of the foam is shown in Fig. 4.

Once preliminary efforts were made to address immediate safety hazards and provide weather protection, a more detailed investigation was performed utilizing visual examination, hammer sounding to identify areas of unbonded cement plaster, as well as invasive openings in select areas to examine the condition of the hollow-core clay block infill back-up walls. This investigation was initiated by the nonprofit owner, at their expense, and was conducted by their consultant. Invasive openings in the exterior wall were also used to verify the results of sounding. Accessing the wall from a boson's chair and swing stage scaffolding, the potential areas of unbonded cement plaster, damage to the hollow-core clay blocks, and cracks in the stucco surface were identified, marked, and mapped onto the exterior surfaces of the walls, as shown in Fig. 5. CAD drawings were prepared to document and quantify the areas of damage. Visual inspection (FEMA Method NDE 1), sounding (FEMA Method NDE 2), and selective removal of materials (FEMA Method IT 1) were used to identify examples of FEMA-recognized damage types, such as corner crushing, diagonal cracking, bed joint sliding, and out-of-plane failure [13].

As outlined above, the cost of the inspection and documentation work needed to identify the types and extent of earthquake damage was not eligible for reimbursement. Therefore, within its limited budget, SCIDPDA directed its consultant to perform only select swing stage drops at each elevation, and extrapolate the average amounts of damage for each elevation. Although not reimbursed, this inspection work proved valuable. Prior to SCIDPDA's independent evaluation, early in the process, FEMA had estimated repairs at approximately $200 000 based on their initial inspection, and the identification of obvious exterior cracking and cement plaster bulges. However, by the start of the repair project, due to careful documentation and appeals, the Project Worksheet had been revised by FEMA to a cost estimate of $1.3 million.

FIG. 4—*Application of spray foam into open void in exterior wall.*

From close examination of the exterior walls, the Project Team was convinced that a greater amount of earthquake-caused damage existed, not easily seen from the ground or from visual observation alone. Identifications of exterior cracking and the obvious bulging as the only damage was too limited to sufficiently describe the damaged incurred by the Building. FEMA publication FEMA 306, *Evaluation of Earthquake Damaged Concrete and Masonry Wall Buildings, Basic Procedures Manual*, states, "...the significance of cracks and other signs of damage, with respect to the future performance of a building, depend on the mode of behavior of the components in which the damage is observed" [14]. FEMA publication FEMA 308, *Repair of Earthquake Damaged Concrete and Masonry Buildings*, points out that cracking can be an unpredictable indicator of damage [15].

The detailed survey and resultant estimated quantities of not only cracking, but also more hidden unbonded cement plaster, crushed and damaged hollow-core clay block, and cracked concrete, were used successfully to negotiate larger repair quantities and a significantly higher cost estimate in the Project Worksheet before putting the project out to bid. The documentation provided in the independent survey was used in several appeals with FEMA, to protest repair estimates that were believed to be too low.

The key to successfully navigating a repair project through the regulatory requirements of FEMA depends largely on thorough documentation. It is important to verify and document conditions encountered directly after the event as well as maintaining diligent recording of conditions throughout the entire repair

FIG. 5—*Cracks, bulges, and areas of unbonded cement plaster were marked on the on the wall. Invasive openings were made to help verify observations.*

process. Establishing standard testing and documentation protocols is necessary to conduct an accurate evaluation with precise measurements that can be clearly recorded and depicted in accompanying repair documents. Since this repair project was exterior wall-related, utilizing elevation drawings to document original damage, additional found damage, and track the progress of repairs was crucial for FEMA representatives to verify and approve work performed on-site as well as authorize payment for those repairs.

Ultimately the information gathered in the independent survey helped to:

- Provide more accurate damage estimates for bid documents (a cost estimate of $1.3 million dollars was established prior to bidding);
- Facilitate better mutual understanding of damage types and damage extent between FEMA and State officials, and the Project Team;
- Lay the foundation and framework for the repair project, minimizing the chance for previously unknown "hidden damages" to disrupt the schedule.

The types of damage that were identified and included in the Project Worksheet are outlined below:

1. Cement Plaster Parge Coat
   a. Heavily fractured and unbonded sections of cement plaster. Unbonded stucco apparently occurred as a result of crushed and cracked hollow-core clay block, as well as movement of the concrete structural frame.
   b. Sections of cement plaster were noted to have fallen from the building during the earthquake.
   c. Hammer sounding at select areas of the north and south elevations indicated that large areas of unbonded cement plaster remained "hanging" at exterior walls, creating potential hazards.

   d. Wide-spread cracking of the cement plaster cladding left the wall assembly vulnerable to water intrusion and freeze-thaw cycling, which could lead to additional cement plaster falling from the building.
   e. The existing elastomeric coating applied to the cement plaster was determined to no longer be of serviceable condition as a result of the extensive cracking in the underlying cement plaster stucco.
2. Hollow-Core Clay Block: The hollow-core clay block is not reinforced, nor cell-grouted:
   a. In numerous locations, hollow-core clay block located at the corners of infill panels, at intersections of the concrete structural frame, was fractured/crushed due to movement of the building's concrete frame during the earthquake. In turn, the damaged clay block negatively impacted the overlying parge coat of cement plaster (i.e., movement, cracking, and crushing of clay block displaced overlying cement plaster). In many instances, the shell and inner webs of hollow-core clay blocks were fractured.
   b. The mortar between units of hollow-core clay block was in poor condition, loose and quite friable in numerous locations indicating weakened infill panels. In several locations, mortar between the hollow-core clay block and concrete frame had cracked, crushed, and separated, or a combination thereof, leaving the clay block infill panels essentially "freestanding," i.e., no longer positively engaged with the concrete frame.
3. Reinforced Concrete Frame: The reinforced cast-in-place concrete skeleton frame structure did not appear to have significant seismically induced damage, i.e., in those places where the concrete structure was exposed, no significant cracking and spalling, or both, of the concrete was observed. However, several limited locations of spalled concrete and exposed steel reinforcement were observed. (Later, during the repair project, additional cracks and spalling were discovered.)
4. Mural at South Elevation: A mural covered approximately 1650 square feet (153 m^2) of the south elevation wall. Deficiencies and damage of the hollow-core clay block substrate, cement plaster parge coat, and consequential deterioration of the overlying paint necessitated repair of the mural.

With the open voids in exterior walls temporarily weather-tight, and the preliminary analysis complete, the Project Team developed repair procedures and bid documents, while negotiating the qualified repairs and cost estimates with Washington State Military Department Emergency Management Division and FEMA representatives.

Repair Concepts

As planning of repair procedures began, the Project Team considered a number of factors. The Bush Hotel, containing retail and office space on the street and second levels, respectively, as well as four floors of housing above, remained fully occupied throughout the duration of the investigation and repairs. Protection of the retail and office tenants, residents, and the public were of paramount concern. It was deemed that the housing service of the nonprofit organization could not be interrupted, thus, repair scenarios were devised that would not impact the interior of housing units and would not require relocation of tenants. (This ruled out, for example, complete replacement of hollow-core clay block with new clay or concrete block, due to the negative impact on residents.) Repair of the building to its pre-disaster state was also considered in context of principles outlined in the publication, FEMA 308, *Repair of Earthquake Damaged Concrete and Masonry Buildings* [16]. Unfortunately, this also prevented consideration of comprehensive retrofit options such as the complete replacement of problematic hollow-core clay block infill at exterior walls with light-gauge steel framing, or over-cladding the entire building.

Historical Restoration Considerations

To ensure the historic status and fabric of the Bush Hotel was not jeopardized, all proposed repairs were carefully scrutinized and planned for their appropriateness (i.e., to help ensure they were historically sensitive). All repair work was designed and performed pursuant to the Secretary of the Interior's "Standards for the Treatment of Historic Properties." The State and FEMA also coordinated review of the Project Worksheet, as well as bid documents, with the State Historical Preservation Office. In addition, the Project Team achieved mutual agreement for repair methods with the local, City of Seattle International District Review Board.

As an example of one of the issues that was discussed between the State and the Project Team was the repair material planned for replacement of damaged hollow-core clay block. The Project Team proposed to fill-in the fractured clay block wall areas with reinforced Type-N mortar, which although was somewhat of a compromise, was selected due to a number of factors.

The National Park Service (NPS), in *Preservation Brief 16: The Use of Substitute Materials on Historic Buildings*, warns that substitute materials should be avoided, when possible, in the repair of historic buildings. However, the document also states,

"In limited circumstances substitute materials that imitate historic materials may be used if the appearance and properties of the historic materials can be matched closely and no damage to the remaining historic fabric will result [17]."

In the context of the Bush Hotel Repair Project it was mutually agreed by all parties that openings in the hollow-core clay block infill panels would be filled-in with reinforced mortar. This decision was made base on the following considerations:

- New hollow-core clay block was relatively difficult to procure, and has a long lead time for the manufacturer of new block;
- There was an unknown quantity of block needed, and if partway through the project more block was needed that was originally anticipated, the resultant delay in procuring more block could have negative scheduling impacts;
- The multi-wythe construction of the existing block walls and the different interlocking shapes of existing blocks made replacement with new block relatively complex. In addition, replacement with new block would require disruption of interior finishes;
- It was deemed that disruption of interior finishes, relocation of tenants, and disruption of the nonprofit organization's services must be avoided;
- It was agreed that substituting mortar in-fill for new hollow-core clay block would not significantly alter the historic fabric of the building, and would not negatively affect the finished appearance of the exterior walls;
- It was agreed that the materials were sufficiently compatible.

Other alternate repair methods and materials, including construction of the infill walls with concrete masonry units (CMU), were considered during the evaluation and design process. The desire to keep the housing facility open and fully functional throughout the entire repair process, along with the difficulty of replacing damaged interlocking blocks without disrupting interior finishes (which were coated with led paint in locations) lead to the solution of repairing the damaged blocks with reinforced mortar infill. FEMA regulations also stipulate that to receive funding for repairs, the subject property is limited to repairs that restore the property to pre-disaster conditions. Providing upgrades to the facility, even if the owner agreed to pay for the additional costs, could jeopardize the funding package provided by FEMA.

Another example of a technical issue considered in light of historical qualities of the building, was the evaluation of the incorporation of lath into the cement plaster patches. Although NPS *Preservation Brief 22: The Preservation and Repair of Historic Stucco* discourages the use of metal lath over masonry substrates [18], stainless steel lath and stainless steel lath fasteners were utilized in the portland cement plaster repair. It was determined by the Project Team that lath would help ensure that patches remained attached to the building during a potential future seismic event, and would help to bridge and reinforce between mortar joints of the hollow-core clay block and joints between block and concrete, thereby reducing cracking. Stainless steel screw fasteners were placed in pre-drilled holes in the clay block and mortar, and stainless steel expansion pins were used in the concrete to fasten the lath to the wall. Both types of anchors proved to have excellent withdraw resistance and did not fracture the hollow-clay block during withdraw testing.

Exterior Mural

As previously stated, a large mural existed at the south elevation of the building (Fig. 6). Because the mural was not part of the original construction (it was painted at a much later date), and the mural was not listed in the Bush Hotel's description in the National Registry, the State Historical Preservation officer determined that it was not "officially" a historical element of the building. All parties agreed, however, that the mural was culturally significant to the community. Therefore, it was determined by the Project Team that the underlying substrate would be fully repaired, and the mural would be repainted. Full repainting of

FIG. 6—*Completed repairs at south elevation of Bush Hotel, including the finished mural.*

the mural, as opposed to restoration of the existing paint, was allowed due to the mural's status. Covering approximately 1650 square feet in area (153 m^2), the mural was carefully documented with scaled photographs prior to the start of repairs. Repairs were completed to the underlying exterior wall assembly, and the stucco substrate wall fully coated with an elastomeric coating to provide a sound base for the mural. The original small oil painting, serving as the model to create the mural was in the building owners' possession, further providing the current artist with a means to scale and match the composition of the original mural. Funding for restoration of the mural was included within the FEMA approved Project Worksheet.

Scope of Repairs

The following outlines the general scope of earthquake damage repair work incorporated into the Project Worksheet for the Bush Hotel (the scope is shown abbreviated here):

1. Provide Access and Protection
2. Remove Unbonded Cement Plaster
3. Remove and Replace Broken Hollow-Core Clay Block
4. Rout and Point Exposed Clay Block Mortar Joints
5. Epoxy Inject Crack(s)/ Patch Damaged Concrete Structural Frame
6. Rout and V-Cut Large Cracks in Cement Plaster
7. Prepare Exterior Walls for Coating
8. Retrofit of Stainless Steel, Helical Dry-Fit Pins
9. Repair/Patch Cement Plaster Cladding, Including Cracks
10. Apply Cementitious Finish Coat
11. Coat Exterior Walls
12. Repaint Mural and Painted Signs

Because of the densely-populated, urban location of the Bush Hotel, and its function serving a residential population, hazard mitigation for future seismic events became a major feature of the repair project. In order to stabilize the hollow-core clay block infill walls, the infill walls were anchored to the structure using stainless steel, helical dry-fit pins installed around the perimeter of each panel. The hazard mitigation measure was prudent in order to: (1) limit the likelihood of damage to the cement plaster during a future minimal-to-moderate seismic event; and (2) to decrease the possibility of complete loss or collapse of the infill panels, to help protect the life-safety of the occupants and the public during a future seismic event. Time and budget constraints influenced the design and implementation of the helical dry-fit pins based on best practice and experience with other similar buildings.

Contract Documents, and Procurement

With approval of the scope of work and cost estimate by FEMA, detailed technical specifications, detail drawings, and general conditions were assembled for the repair project. Due to the public nonprofit status of the owner, and in light of FEMA requirements, it was decided that the project would be publicly and competitively bid. Thus, the Project Team focused on producing comprehensive and detailed contract documents. Important aspects of the Contract Documents, and the bidding process included:

1. *Unit Cost Pricing*: The Bid Form included a detailed line-itemized scope of work, with estimated quantities for each work item, and required the bidding contractor to fill-in the unit cost for each item. The unit costs were inclusive of overhead and profit. The contract amount was determined by multiplying the unit costs by the stated quantities, and providing the sum total (not inclusive of bonds, insurance, and tax.). The contract stated that the unit costs could be used in additive fashion during the duration of the work, based on the quantities of work in each category actually performed. This way, there was a definitive contract price established for each bidder, but at the same time, there was good control provided to the Project Team to adjust the contract amount based on the work performed. This became a very important attribute in the Bush Hotel project, as damage quantities, and thus the cost of the project, more than doubled as additional hidden damage was discovered. It was very important for the Project Team and FEMA to have a previously agreed upon mechanism to increase the contract amount.
2. *Selective Demolition and Hold Points*: Throughout the early part of the project, FEMA and the Project Team fundamentally disagreed about the extent of damage. The Project Team maintained that there were extensive areas of crushed hollow-core clay block that were unseen, and which could not be fully quantified prior to the repairs. However, FEMA maintained that per regulations, only evident damage could be included in the Project Worksheet estimate. This was a point of conflict that had to be addressed. Typically, as stated above, repairs can only be made on approved quantities, and initially, only visible cracks and bulges were verified as damage by FEMA. However, it was finally established that the Repair Contractor be allowed to conduct selective demolition of all unbonded cement plaster and damaged clay block at each individual elevation, *without* the quantity of the selective demolition being established for that elevation prior to the start of demolition. This agreement was crucial to ensuring the success of the Bush Hotel project. Without this agreement, the Project Team was potentially at risk for jeopardizing the funding by performing a greater quantity of selective demolition than allowed in the Project Worksheet.

   The Contract Documents were written such that the contractor completed selective demolition of unbonded cement plaster parge coat, and damaged clay block, at one elevation at a time. The selective demolition was spot-checked monitored to help ensure that only damaged material was removed. Once selective demolition was complete, the contractor was required to hold their work at that elevation, while the contractor and consultant worked together from a swing stage to document the extent of damage. FEMA then verified calculations of the extent of damage, which became the undisputed quantity of damage for that elevation. Once verified, the contractor was allowed to begin repair work. The FEMA verified quantities then became the basis for change orders.
3. *Breakdown of Repair Quantities By Building Elevation*: Estimated damage and repair quantities were listed in the Project Documents by each individual exterior wall elevation. This allowed the contractor to stage various different types of work at each elevation, and to avoid scheduling delays that may have otherwise been caused by the FEMA verification process. For example, while verification of the quantity of cement plaster and clay block that had been removed occurred at one elevation, repairs could be ongoing at another. Without this provision, the contractor would have had to complete selective demolition at all elevations, and the project would have then been shut down to wait for FEMA verification.
4. *Documentation*: It was made clear that all repair quantities must be explicitly documented, and that the contractor plan for the time necessary for the documentation and verification to take place at each elevation.
5. *Anticipation of Hidden Damages, and Other "Protective" Measures*: The language of the Contract Documents was crafted to "protect" the administrative process required to maintain and ensure funding through the Public Assistance Program. For example, the steps for identification of hidden

damage, documentation, verification, approval of the increase in work and cost, and the time the contractor would be required to wait (without opportunity for delay claim), were explicitly outlined.

6. *Qualification Process*: As part of the bidding process, bidding contractors were required to submit a detailed proposal outlining their qualifications, in addition to their bid proposal. Packages were given to each interested general contractor, which included detailed questionnaires, including listing the qualifications required of important subcontractors such as the plastering and masonry trades. The qualification packages clearly indicated the scoring process. For the Bush Hotel project, the qualification process helped to identify qualified and competent contractors from the bidding pool, even in a public bid situation.

As required, the Project Contract Documents were submitted to the State and FEMA for review, prior to bidding. The Project Documents underwent a multi-agency review process, including review of historical preservation and environmental impacts. Once the Project Documents were approved, they became the ground rules for interaction between FEMA, the State, the Project Team, and Repair Contractor. Written in a detailed and specific fashion for the Bush Hotel project, the documents helped to eliminate much of the confusion that could have otherwise occurred due to the administrative demands of the project.

Repair Project

As previously discussed, the building remained fully occupied during the duration of the investigation and repair process. Careful coordination between the Project Team, including the consultants, contractor, owner, and owner's representative was crucial. Additional watertight scaffolding canopies were provided at entries, and protective fencing and canopies were installed at the south elevation adjacent to an active city park, to protect the public and those using the building.

Repairs began at the east elevation and continued around the building in a counter-clockwise fashion, sequentially moving to the north, west, lightwells, and south elevation; allowing the contractor to operate several crews progressing through the various stages of repair at different elevations simultaneously. Utilizing swing stage scaffolding to access the wall cladding surface, the cement plaster cladding was first washed with hot pressurized water to remove the majority of the existing elastomeric coating, which was in poor condition. (This was considered part of selective demolition, better exposing and preparing cracks for repair.) As part of the safety procedures, paint chips and water run-off collected during the washing process were tested for lead. Lead was present at unacceptable levels requiring paint chips to be collected and stored in protective drums for proper removal and disposal from the site. Run-off water from the pressure-washing process was also collected and filtered prior to being properly discharged.

As selective demolition continued, sound testing was conducted by the contractor to determine the extent of unbonded cement plaster parge coat. Once identified, loose plaster was removed. At locations of unbonded cement plaster over hollow-core clay blocks, the bond between the cement plaster and the outer surface of the blocks was often intact. It became apparent that in many situations, the weakest juncture of the exterior wall cladding assembly sustaining the most damage was the hollow-core clay block web itself. Examination of the debris during selective demolition revealed that while the bond between the cement plaster stucco and outer surface of the clay block remained adhered, the matrix of the hollow-core clay block itself was often destroyed, necessitating the need to remove large portions of fractured block pieces from the wall cladding assembly at several locations. Other samples of unbonded cement plaster revealed that only the thin outer surface of the hollow-core clay block experienced delamination, sheared just beneath the exterior surface, leaving the remaining portion of the block intact, as shown in Fig. 7.

Individual areas of unbonded cement plaster cladding were generally larger in size and more extensive at the lower floors of the building. Photos of the north elevation illustrate the greatest concentration of damage over the cast-in-place concrete structure of the lower floors, gradually experiencing less damage at the upper portions of the elevation (Fig. 8). Of the estimated 38 813 square feet (3606 m^2) of exterior wall surface, approximately 7467 (694 m^2), or approximately 19.2 % of cement plaster cladding was removed and replaced.

Cracks in the cement plaster were repaired. Cracks measuring approximately 1/32-in. or greater were mechanically routed into a v-cut groove, cleaned, and repaired with cement plaster finish coat. Cracks measuring less than approximately 1/32-in. in width were cleaned of debris to receive a stripe coat of

FIG. 7—*Unbonded cement plaster and shearing of clay block matrix.*

elastomeric coating during the coating process. While some hairlines cracks appeared as only superficial deterioration, close examination of the cement plaster matrix following removal of the existing elastomeric coating revealed that even the thinnest of cracks had propagated through the entire depth of the parge coat.

Throughout the entire selective demolition process, accurate quantities of all damage and resultant repair items were measured and documented for the Washington State Emergency Management Division and FEMA's verification and approval. In the preceding days leading up to on-site quantity verification by the State or FEMA, the consultant and repair contractor collaborated to survey the exterior wall surface. Detailed spreadsheets and elevation drawings were prepared for each elevation to document the extent of damage and resultant required repairs, as shown in Figs. 9 and 10. The repair quantities required for field verification included: square footage of cement plaster demolition, hollow-core clay block demolition, and concrete repair, as well as length measurements for clay block joint repointing, epoxy injection of cracks in the concrete structure (if needed), and length of cracks in the cement plaster. Providing the Washington State EMD and FEMA representatives with detailed drawings keyed to accompanying spreadsheets sum-

FIG. 8—*North elevation of building after selective demolition, depicting the openings in the exterior wall, and removed sections of unbonded cement plaster.*

Bush Hotel - Earthquake Damage Repair Project

Cladding Demolition Areas — August 4, 2005

NORTH ELEVATION						
Demo Area I.D. Number	Cement Plaster Selective Demo (Sq. Ft.)	Hollow Clay Block Selective Demo (Sq. Ft.)	Hollow Clay Block Joint Pointing (Lin. Ft.)	Concrete Repair (SF) - Patch w/Rebar	Concrete Repair (SF) Patch no Rebar	Concrete Repair (LF) Epoxy Inject
N1-01	2	2.5	0	0	0	
N1-02	3	4	0	0	0	
N1-03	1.5	1.5	0	0	0	
N1-04	8	2.5	4	0	0	
N1-05	3	1.5	3	0	0	
N1-06	4	3	0	0	0	
N1-07	2	3	0	0	0	
N1-08	1.5	2	0	0	0	
N1-09	1	1.5	0	0	0	
N1-10	1.5	0	0	0	0	0.75
N1-11	16	1.5	14	0	0	1
N1-12	1	1.5	0.5	0	0	
N1-13	0.5	1	0	0	0	
N1-14	3.5	2.5	0.5	0	0	
N1-15	2	1.5	0	0	0	0.75
N1-16	3.5	2.5	12	0	0	
N1-17	6	4.5	0	0	0	
N1-18	18	7	11.5	0	0	
N1-19	5	0	4.25	0	0	
N1-20	31.5	16	4	0	8.5	4
N1-21	8	3	2	0	0	
N1-22	20	8	0	0	0	
N1-23z	9.5	0	12	0	0	
N1-23dd	16	4	4	0	0	
N1-23ee	6	4.5	1	0	0	
N1-23ff	9.5	4.5	0	0	0	
N1-23ii	23.5	14.5	13	0	0	
N1-23jj	7.5	3.5	5.5	0	0	
N1-23kk	19	8	1	0	0.5	4
N1-23ll	25.5	12.5	0	0	0	6
N1-23mm	21	5	0	0	0	1
N1-23nn	12.5	4	0	0	0	4
N1-23oo	36	9	4	0	2	3
N1-23pp	21	12.5	12.5	0	0	
N1-23qq	2.5	2.5	0	0	0	
N1-23rr	28.5	13	6	0	0.5	
N1-23ss	13	10.5	4	0	0	
N1-24	28.5	13	6	0	0.5	6
N1-25	3.5	2	2	0	0	
Total:	**425.5**	**193.5**	**126.75**	**0**	**12**	**30.5**

FIG. 9—*Example of detailed spreadsheet, summarizing damage and repair quantities. Locations listed in spreadsheet were keyed to CAD drawing.*

marizing the damage and repair quantities enabled accurate and expedient verification. The documentation will also serve as a valuable record of repairs for the owner.

Once quantities were approved at an individual elevation, repairs could begin. In many locations, because of the relatively large amount of mortar that would be mixed and placed into the open cavities of infill walls, the contractor pumped the mortar to the damaged areas, applying the mortar similarly to spray-applied cement plaster. Typically, prior to placing the mortar within the cavities, 1/8-in. diameter galvanized steel rod was installed in an approximate 6-in. grid to provide additional reinforcement, as shown in Fig. 11. Other elevations required more extensive reinforcement. Cementitious finish coat was applied over repaired areas to match the existing exterior wall texture in locations of cement plaster patches. Several mock-ups were prepared by the contractor, and reviewed by the Project Team, in order to establish a texturing method that closely replicated the existing cement plaster and allowed the new patches to blend well, an important aspect of maintaining the historic fabric of the building.

Where hollow-core clay block infill was exposed and in good condition, mortar joints were routed and pointed with Type-N mortar.

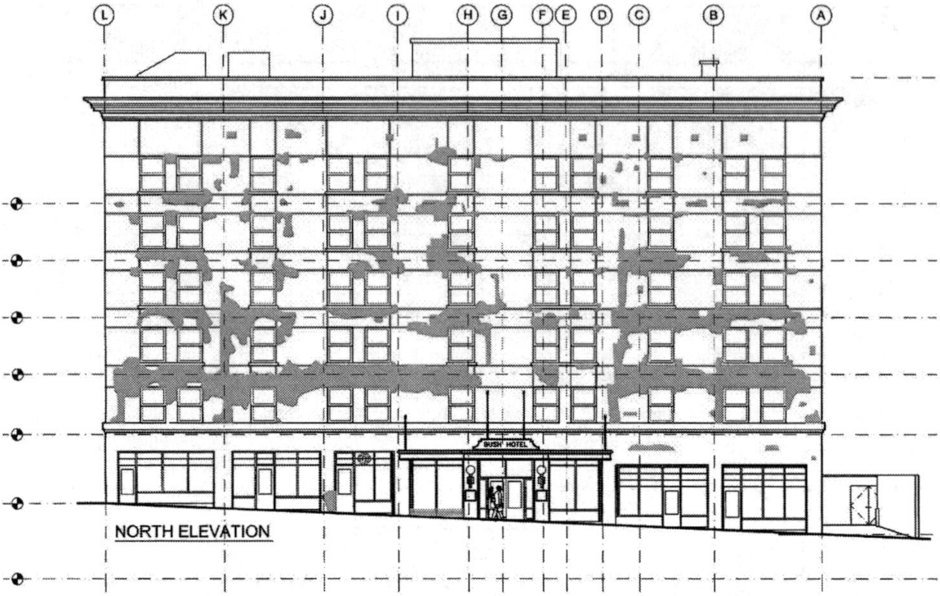

FIG. 10—*CAD drawing illustrating actual areas of damage measured following selective demolition. Each area of damage was mapped, and the quantity tabulated.*

Stainless steel helical dry-fit pins were installed at approximately 2-ft (61-cm) on center around the perimeter of clay block infill panels, as well as at areas that received the mortar repair. As stated, the intent of the pins was to mechanically fasten the infill panels to the concrete frame. Typically the pins would be drilled and installed after the mortar cured. A pin is shown in a photo of the "mock-up repair" Fig. 11.

Following the cure of the mortar infill, stainless steel expanded metal lath was mechanically fastened in cement plaster repair areas. Specifications stipulated that stainless steel lath to be fully embedded in a cement plaster "scratch coat." Application of a "brown coat" followed to thoroughly engage, adhere, and bond to the scratch coat. The cement plaster finish and texture coat was applied to match the existing

FIG. 11—*Mock-up of typical repair. Note helical pin inserted into concrete spandrel.*

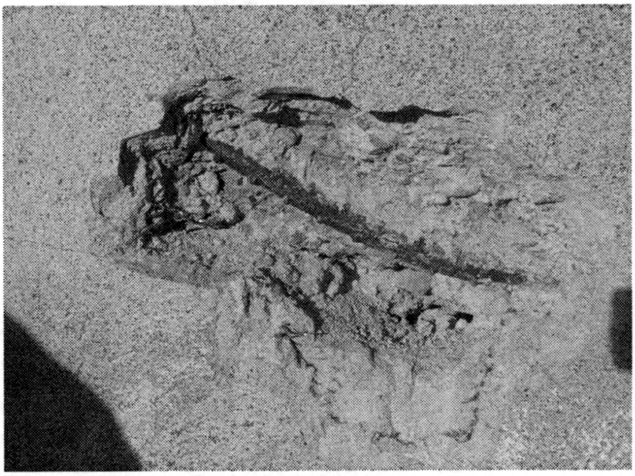

FIG. 12—*Typical localized concrete damage.*

cladding surface adjacent to the repair area. An overall thickness of the new cement plaster cladding typically ranged from 3/4 in. to 1 in. (1.9 to 2.54 cm). Because of the existing conditions and original construction tolerances, matching the thickness of the existing cement plaster cladding required a thinner application in some locations.

Evidence of spalled concrete at select locations were uncovered during the selective demolition, primarily at cold joints, or areas of inadequate coverage of the embedded steel reinforcing, as shown in Fig. 12. In locations of exposed reinforcing, the steel was prepared and painted, the concrete was prepared, and the patch was made. Cracks in the concrete structure were epoxy injected; however, damage to the concrete structure was relatively limited.

The greatest amount of damage was uncovered at the south elevation. Instead of the typical double-wythe hollow-core clay block infill wall, only a single-wythe back-up wall existed at the south elevation, exacerbating the quantity of damage discovered. As selective demolition continued, the Project Team quickly realized that the quantities of damage at the south elevation was much larger than originally expected, prompting further evaluation by the structural engineer. The southern exposure, single-wythe back-up wall, and seismic forces running parallel with the wall plane, all contributed to the relatively extreme damage uncovered at the south elevation, as shown in Fig. 13. The conditions at the south elevation constituted "hidden damage," as the damage significantly exceeded original estimates and expectations, and required additional funds to be allocated for the additional repairs.

For the extensive damage at the south elevation, additional steel reinforcing was specified by the structural engineer, consisting of epoxy-set, No. 4 rebar placed at 32-in. on-center to provide further stabilization of the infill walls. The additional rebar was installed in conjunction with a secondary framework of 1/8-in. diameter galvanized steel rod in a grid pattern. The change in the proposed reinforcing methods and materials was subject to review and approval by FEMA. Approval of the changes also required FEMA to substantially revise the Project Worksheet (PW) to include repair of the "hidden damage," Letters from the structural engineer and consultant were submitted along with photographic documentation of the damage. The importance of planning and close communication with the Washington State EMD and FEMA officials proved critical as repairs could not continue on the south elevation until approval was granted and the revised Project Worksheet was issued. Following a lengthy review process by FEMA, which nearly stalled the repair work and left the cavities of the hollow-core clay block exposed to the elements during the Pacific Northwest's rainy season, approvals were granted and implementation of the additional reinforcing techniques promptly began.

Installation of the epoxy-set, No. 4 rebar by the contractor was performed under the observation of a special inspector through the City of Seattle. The City of Seattle required a special inspector to observe and verify the epoxy-set rebar repairs. This requirement again demanded that all parties carefully coordinate and document the work performed. Specifications from the structural engineer called for the rebar to be

FIG. 13—*Photo illustrated severe damage at south elevation exposed during selective demolition of unbonded cement plaster and fractured clay block.*

installed 6-in. from the corners of the concrete structural frame, as shown in Fig. 14 along with the completed installation of reinforcing, shown in Fig. 15. The additional framework of steel reinforcing bars at the south elevation was provided to stabilize the repair materials within the exterior plane of the structure in the event of future seismic activity. Repairs began at the eastern end of the south elevation and continued to the west, allowing for the start of the mural painting while repairs continued on other parts of the elevation.

Restoration of the mural on the south elevation marked the final component of the earthquake repair project. The restored mural serves as an important cultural icon for the neighborhood and is a source of pride in the community. In addition, other painted signs on adjacent elevations were also repainted.

Conclusions

Although the extent of the damage was larger than first anticipated, causing the Earthquake Repair Project to extend past the original time estimates, as well as increase significantly in cost, clear communication

FIG. 14—*Epoxy-set reinforcing bar at area of significant hollow-core clay block damage.*

FIG. 15—*Overview of reinforcement installed at south elevation in preparation for application of mortar, and new cement plaster cladding.*

helped to ensure a successful project. Continual coordination among the nonprofit organization, consultants, and contractor enabled the Project Team to keep FEMA and State Emergency Management officials informed of the progress and helped the administration of the project to progress smoothly. The State and FEMA representatives were responsive and informative, further helping the Project Team to ensure compliance with the variety of regulations.

A brief summary of several of the most successful aspects of the Bush Hotel project are:

- Technical aspects of the repair, including, matching the plane and texture of cement plaster patches to the existing parge coat, dealing with large areas of fractured hollow-core block, and successfully repainting the mural to match the original.
- Implementation of the hazard mitigation plan in an integrated fashion with earthquake damage repairs.
- The independent survey of the damage, including detailed documentation, that laid the foundation for identification and quantification of earthquake damage and necessary repairs.
- The provisions in Contract Documents that explicitly interwove the requirements mandated by FEMA with the contractor's specified work.
- The agreement established between the Project Team and FEMA that selective demolition could be conducted prior to the exact quantity of selective demolition being known. As shown, this helped the project to progress in a timely fashion, allowing all documentation and verification of actual damage following selective demolition.
- Preservation of the historic fabric of the building, and coordinated review of the repair process with State and local historical preservation agencies.
- Completion of the project while facility was in full use, and fully occupied.

When dealing with the myriad of regulatory agencies, review boards, and funding constraints, compromises will need to be negotiated by all parties involved to find an appropriate solution that ultimately serves the best interest of the building. The importance of conducting thorough evaluations early in the project followed by accurate documentation and maintaining open lines of communication throughout the entire process can help produce highly successful results. In the finished product, the neighborhood was rewarded with the historically sensitive restoration of an important community facility.

References

[1] *Washington State Public Assistance Manual*, State of Washington Military Department Emergency Management Division, Olympia, Washington, 2006, p. 3.

[2] FEMA 321: *Public Assistance: Policy Digest*, Federal Emergency Management Agency, Washington, DC, 2001, p. 36.
[3] FEMA 322: *Public Assistance: Public Assistance Guide*, Federal Emergency Management Agency, Washington, DC, 1999, p. 8.
[4] FEMA 321: *Public Assistance: Policy Digest*, p. 59.
[5] FEMA 321: *Public Assistance: Policy Digest*, p. 4.
[6] FEMA 322: *Public Assistance: Public Assistance Guide*, pp. 27–28.
[7] FEMA 322: *Public Assistance: Policy Assistance Guide*, p. 98.
[8] FEMA 322: *Public Assistance: Policy Assistance Guide*, p. 102.
[9] *Washington State Public Assistance Manual*, p. 23.
[10] FEMA 154: *Rapid Visual Screening of Buildings for Potential Seismic Hazards: A Handbook*, Applied Technology Council (ATC-21 project), Federal Emergency Management Agency, Washington, DC, 1988, p. 28.
[11] FEMA 154: *Rapid Visual Screening of Buildings for Potential Seismic Hazards: A Handbook*, Applied Technology Council (ATC-21 project), Federal Emergency Management Agency, Washington D.C., 1988, p. 27.
[12] FEMA 306: *Evaluation of Earthquake Damaged Concrete and Masonry Wall Buildings*, Applied Technology Council (ATC-43 project), Federal Emergency Management Agency, Washington, DC, 1999, p. 188.
[13] FEMA 306: *Evaluation of Earthquake Damaged Concrete and Masonry Wall Buildings*, Applied Technology Council (ATC-43 project), Federal Emergency Management Agency, Washington, DC, 1999, pp. 50, 52, 54, 206–210.
[14] FEMA 306: *Evaluation of Earthquake Damaged Concrete and Masonry Wall Buildings*, Applied Technology Council (ATC-43 project), Federal Emergency Management Agency, Washington, DC, 1999, p. xvii.
[15] FEMA 308: *Repair of Earthquake Damaged Concrete and Masonry Wall Buildings*, Applied Technology Council, Federal Emergency Management Agency, Washington, DC, 1999, p. 9.
[16] FEMA 308: *Repair of Earthquake Damaged Concrete and Masonry Buildings*, Applied Technology Council, Federal Emergency Management Agency, Washington, DC, 1999, p. 9.
[17] Park, S. C., AIA, *Preservation Briefs 16: The Use of Substitute Materials on Historic Building Exteriors*, U.S. Department of the Interior National Park Service Preservation Assistance Division, Washington, DC, 1988, p. 1.
[18] Grimmer, A., *Preservation Briefs 22: The Preservation and Repair of Historic Stucco*, U.S. Department of the Interior National Park Service Preservation Assistance Division, Washington, DC, 1990, p. 10.

EXTERIOR INSULATION AND FINISH SYSTEMS (EIFS)

Journal of ASTM International, Vol. 5, No. 4
Paper ID JAI100874
Available online at www.astm.org

Stephen S. Ruggiero[1] *and Annemarie L. Rabazzi*[1]

Design of Engineered Drainable Exterior Insulation and Finish System (EIFS) Cladding

ABSTRACT: During the last ten years, the authors have investigated a number of buildings with leakage and deterioration problems associated with exterior walls clad with exterior insulation and finish systems (EIFS). These systems were "barrier" walls applied to materials that degrade rapidly from exposure to moisture, such as gypsum board sheathing. These wall systems lacked backup waterproofing and flashing elements and relied on surface seals as the sole means of controlling water penetration (leakage). While these systems are appealing due to their aesthetic features, light weight and insulating characteristics, they are vulnerable to performance problems as a result of failed surface seals, leakage through window frame corners and poor installation practices. Installation of internal flashings is difficult at best because the primary waterproofing layer (the lamina) is on the face of the cladding rather than in the plane of the flashing. In this paper, the authors review the essential differences between barrier and drainable EIFS. The primary focus of the paper is to examine the various components of an engineered drainable EIFS including waterproofing membranes, drainage plane grids, methods of attachment, and consideration for use of a vapor retarder (position) within the wall cross section. Detailing of wall penetrations, particularly for windows and doors will be presented as will location of flashings and weep mechanisms in multi story construction. Design considerations will be illustrated in two case studies where barrier EIFS was replaced with an engineered drainable system, one in a relatively warm climate (southern California) and the other in a relatively cold climate (Massachusetts).

KEYWORDS: barrier EIFS, drainable EIFS, water-resistive-membranes, flashings, window openings, head flashing, sill flashing, drainage grid, floor line weeps, vapor retarder, thermal-moisture analysis, WUFI computer model, vapor drive analysis

Introduction

Exterior Insulation and Finish System (EIFS) has been used in the United States to clad exterior walls since the late 1960s. Until recently, an EIFS was typically designed as a surface-sealed barrier wall adhered to the exterior wall sheathing. The barrier EIFS relies on the exterior surface (lamina) of the system and sealant joints at wall penetrations and transitions to protect the backup wall and interior finishes from moisture. Installation of effective flashing details at vulnerable locations in the wall (e.g., windows, doors, vents, and decks) is difficult at best because the flashing inside the wall must return to, and be integrated with, the surface barrier on the face of the wall. In the late 1990s, investigations of barrier EIFS in North Carolina and elsewhere found significant moisture-related damage to building framing components [1]. In response to these problems, manufacturers and designers developed a drainable EIFS that incorporates an internal barrier and drainage plane to collect intruding water and direct it back to the exterior.

Components of a drainable EIFS are similar to those of traditional stucco systems and include a water-resistive barrier over the exterior sheathing; a drainage plane; and in the case of EIFS a cladding material of EPS insulation board, layers of base coat, reinforcing mesh, and a finish coat (Fig. 1). The water-resistive barrier and drainage plane protect the exterior wall sheathing from moisture that may penetrate the lamina at cracks, sealant failures, or improperly flashed wall penetrations and window openings. The internal water-resistive barrier increases the reliability of the system against leakage and facilitates the design and construction of flashings, which can drain internally at the drainage plane rather than returning to the face of the cladding. A drainable EIFS is ideally suited as a replacement or renovation cladding for a failed EIF barrier system as the drainable EIFS is still lightweight and uses the same finish

Manuscript received October 24, 2006; accepted for publication March 21, 2008; published online April 2008.

[1] Senior Principal and Senior Engineer, respectively, Simpson Gumpertz & Heger Inc., 41 Seyon Street, Building #1, Suite 500, Waltham, MA 02453.

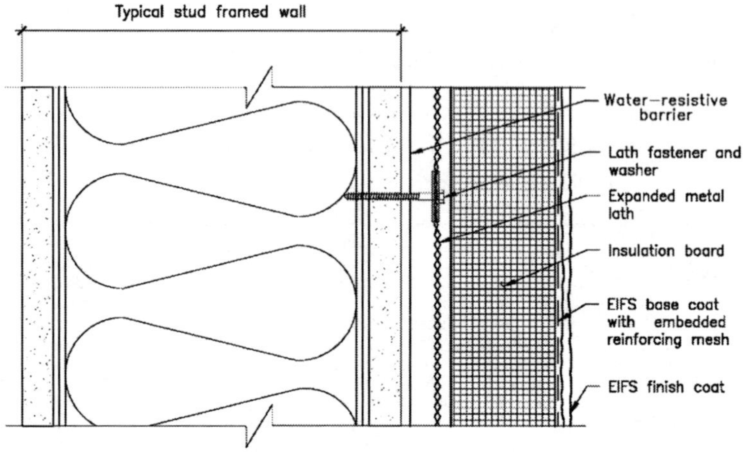

FIG. 1—*Engineered drainable EIFS cladding on stud framed wall.*

material. This last point is often critical in residential condominium structures where a change in aesthetics is likely to be controversial.

In this paper, we will examine the various components of an engineered drainable EIFS cladding and compare it to typical off-the-shelf systems from manufacturers. Design considerations include water-resistive barrier type, drainage plane material and geometry, insulation attachment, vapor drive analysis, and design and location of flashings. These design considerations will be explored in two case studies where barrier EIFS was replaced with an engineered drainable system—one in a relatively warm climate and the other in a relatively cold climate.

Engineered Drainable EIFS

Manufacturers of EIFS wall claddings provide "off-the-shelf" drainable systems with proprietary components that include the type of water-resistive barrier, configuration of the drainage plane, methods of attachment, and guidelines for detailing wall penetrations and flashings. Most manufacturers provide systems that can be engineered or designed to meet the needs of a given project. Engineered drainable EIF systems utilize the manufacturer's cladding (insulation and lamina) in conjunction with drainage materials and a water-resistive barrier either from the EIFS manufacturer or provided by others. The following provides our evaluation of typical system components and ways to increase system performance by utilizing alternative materials.

Water-Resistive Barrier

The backup water-resistive barrier for typical off-the-shelf systems is a liquid-applied coating, usually acrylic or cementitious based, that is applied directly to the sheathing. Joints in the sheathing are stripped with a flexible self-adhered tape or reinforced with a glass-fiber mesh applied with the water-resistive barrier coating. These water-resistive barriers as yet do not have a proven long-term track record, and we are concerned with their performance in a moist environment. The window bays within a wall facade are the most vulnerable to leakage, as opposed to the field of the wall (solid cladding), and warrant consideration for a proven water-resistive barrier and flashing system. In general, the liquid-applied water-resistive barrier is a "field-fabricated membrane" and susceptible to potential installation problems including proper substrate preparation, mixing, and application thickness.

A drainable EIFS gives designers increased flexibility in choosing a water-resistive barrier because the system has dual insulation layers (within the stud cavity and the cladding). Low-permeance membranes such as a self-adhered rubberized asphalt/polyethylene membrane may be an option even in colder climates by balancing the amount of insulation used in both layers to ensure the membrane stays above the dewpoint temperature of the occupied space. The ability to use a self-adhered rubberized asphalt/ polyethylene membrane has its advantages which include the following:

- The membrane is self-sealing around fasteners.
- Membrane is a more-positive installation when counterflashing (overlapping) metal flashings at window heads or weep lines, and it can be formed into sill pan flashings at windows.
- Use of a product with a long-term track record.
- A sheet-applied membrane can bridge small gaps in the sheathing.
- The quality and thickness are not applicator dependent.

Products such as asphalt-impregnated felts used in multiple layers are also an option and have a long-term track record of use in traditional stucco systems. Asphalt-impregnated felts are not as versatile as self-adhered membranes, but are a good option when the water-resistive barrier needs to be permeable to vapor. As will be illustrated later, they can be used in the field of the wall in cold climates to preclude condensation for the general wall assembly and self-adhered membrane can be used in the "window bays" to provide additional protection against leakage.

Drainage Plane

A true drainage plane between the water-resistive barrier and insulation board is important in order to allow the system to dry and avoid a build-up of hydrostatic forces that may allow water to penetrate the water-resistive barrier at fasteners, sheathing joints, or imperfections in the installation. Methods for constructing the drainage plane include the following:

- Insulation board adhered to the water-resistive barrier with vertical ribbons of adhesive.
- Channeled (notched) insulation board on the blind side, attached with either vertical ribbons of adhesive or mechanical fasteners.
- Drainage mat between the water-resistive barrier and the mechanically fastened insulation board.
- Metal lath mechanically fastened over the water-resistive barrier and insulation board adhered to the lath.

A drainage plane that is formed with vertical ribbons of adhesive provides little space to drain and for air to circulate behind the insulation board. The insulation board must be pressed firmly to the substrate to achieve good adhesion, which in turn flattens the ribbons and leaves little air space.

Channeled insulation boards typically have small notches spaced 30.5 cm (12 in.) on center. The channels are difficult to align between boards, and if not aligned correctly will provide a discontinuous path for water to drain within the system. Fasteners used to attach the insulation may be subjected to hydrostatic forces if the drainage plane sees a high water demand.

Use of a drainage mat or metal lath placed between the insulation and water-resistive barrier has the advantage of providing a continuous open space for prompt drainage. Fasteners used to attach the mat or lath will penetrate the water-resistive barrier, which is a disadvantage to this approach that is best mitigated by using a self-sealing membrane for the water-resistive barrier. Adhering the insulation board to the lath has advantages over other adhered systems because the lath holds the insulation board off of the water-resistive barrier so that the adhesive will not fully compress on the wall, and the strength of the cladding (due to negative wind load) is not dependent on the bond between water-resistive barrier and insulation. Care must be taken not to clog the lath with a heavy application of the adhesive.

Insulation Attachment

Most manufacturers recommend adhesive attachment of the insulation boards rather than mechanically fastening to the substrate. In an adhered system, the wind load is distributed over the entire insulation board instead of only at the fasteners. Also, the adhered system limits thermal bridging through the fasteners between the backup wall and lamina.

We see an advantage to using both adhesive and mechanical fasteners to secure the insulation board, particularly at salient corners where the wind loads are greatest. The mechanical fasteners supplement the adhesive attachment at insulation boards that are most likely to fail under high negative wind loads. When using metal lath as the drainage plane, narrow insulation boards that cannot be adhered over multiple lath fasteners should be mechanically fastened to reduce deflection under negative wind loads.

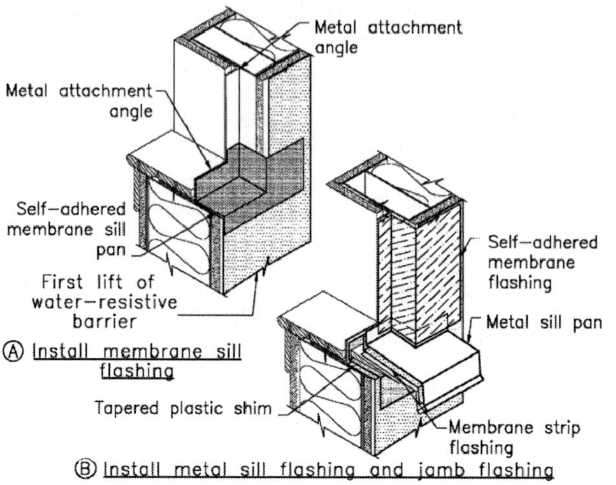

FIG. 2—*Installation procedure for membrane and metal sill pan flashing for a drainable EIFS.*

Basic Design Details

Windows—As with all wall-cladding systems, detailing at window openings is critical to wall performance. While most manufacturers provide details for flashing window heads and sills, these flashings must be sealed or lapped onto the liquid-applied membrane on the surrounding wall sheathing. This results in flashings that are mis-shingled to the flow of water (e.g., flashing at window head). Fishmouths or gaps in the flashing are common installation defects in self-adhered tapes and membranes and provide avenues of moisture penetration that can debond the flashing and create a leakage path over time. For this reason as well as others, we prefer, as a starting point for waterproofing design, to use a self-adhered membrane within the window bays of the facade and to install the sheets from the bottom of the wall to the top in shingled fashion to the flow of water.

Self-adhered sill flashings have been used increasingly in a number of cladding systems, replacing sheet metal flashings that are more complicated and often require another trade to coordinate with the water-resistive barrier installation (Fig. 2). We have used them successfully on a number of projects and design them to drain directly into the wall cavity or drainage plane, which greatly simplifies their construction. Metal flashings are still preferred for use at window heads to protect this vulnerable location and to serve as a weep location for the drainage plane. Most membrane flashings cannot be used to form an exposed drip edge and should be used in conjunction with a metal flashing with an exposed drip edge to drain water from the wall.

Flashing at Floor Lines (Weeps)—In multi-story construction, we prefer to weep the drainage plane or wall cavity at each floor line to limit the flow of water within the system and reduce the risk of leakage through the water-resistive barrier. The location of the flashing/weep assembly should allow for long horizontal runs without interruption by other facade elements, thus simplifying installation details. For most high rises, two choices for flashing locations exist:

- One choice is a line coincident with the top of windows to serve the dual function of weeping the floor above, while protecting the window head. This can be easily implemented if the windows are flush with the face of wall. If the windows are set back from the face of wall, the flashing will need to jog (turn 90°) at each side of the window opening, complicating its construction significantly.
- The other choice is to locate a flashing within the horizontal control joint at floor lines. These joints are required within the cladding to accommodate inter-story drift (lateral deflection), and they create a shadow line that hides the exterior edge of the metal flashing.

Vapor Drive Analysis

Water vapor can migrate through building walls in two ways: by diffusion through the component materials and by direct transport when moisture-laden air flows through gaps (openings) in the wall components. Vapor diffusion occurs on a molecular level and requires a sustained vapor pressure differential across the wall assembly. To address this mode of transport, a vapor retarder should be considered for use within the wall assembly. This retarder needs to be positioned correctly within the wall (as discussed later), but does not need to be perfectly sealed at penetrations and interruptions (e.g., floors and ceilings). Moisture transport within air flows occurs on a macroscopic level and can quickly deposit relatively large accumulations of moisture on surfaces within the air stream that are below the air's dewpoint temperature. To address this mode of transport, an air barrier that can resist wind pressures must be installed within the wall system and it must be sealed completely at all penetrations and interruptions. Even small discontinuities in the air barrier can significantly degrade its performance.

The American Society of Heating, Refrigeration and Air-Conditioning Engineers, Inc. (ASHRAE), divides the United States into three general climate zones: heating, cooling, or mixed. Heating climates are cold weather areas where the interior is heated most of the year and designers are generally concerned with vapor migrating from the interior into the cold outer wall components where it may condense. Cooling climates are warm, humid areas, typically in the southeastern United States, where the interior is air conditioned (cooled) most of the year and designers should generally be concerned with vapor migrating from the exterior to the cold interior components. Mixed climates are located in the central United States and generally do not have a strong sustained vapor drive in one direction.

Many building codes now have requirements for thermal, air, water, and vapor "barriers" within walls to enhance energy efficiency and to preclude hidden condensation within wall systems. Designers have a number of materials to construct these "barriers" and must determine their proper selection and placement within the wall construction. The following are points to consider:

- The water-resistive barrier must be placed over the exterior sheathing to protect moisture sensitive substrates from rainwater. Often, this plane within the wall is an ideal location to construct an air barrier because the barrier is not interrupted by floors, columns, or interior partition walls. Also, windows and doors can easily be sealed and integrated with that barrier.
- The vapor retarder generally is placed on the "winter" warm side of the insulation to prevent vapor within the building from migrating into the cold elements of the wall. This makes sense for most heating climates in the northern United States, where moisture generated in the occupied space is contained by a vapor retarder placed between the insulated stud wall and the interior finishes.
- In the southeastern United States, the vapor drive is generally in the opposite direction of northern climates, and therefore, the vapor retarder should be positioned to limit the migration of exterior vapor into the building, where it may condense upon reaching cold air conditioned surfaces. In some respects this can simplify barrier design as the waterproofing membrane, vapor retarder, and air barrier can be constructed from one material placed over the exterior sheathing.
- In northern climates, a drainable EIFS gives designers increased flexibility in wall design because the EPS insulation provides an insulating layer outboard of the water-resistive barrier. By adjusting the thickness (R-value) of the exterior EPS insulation and the interior batt insulation, it is often possible to maintain temperatures in all parts of the backup wall that are warm enough to prevent condensation during winter. This typically requires increasing the thickness of the exterior insulation and decreasing the thickness of the batt insulation within the stud wall while maintaining the overall code prescribed R-value for the wall assembly. If conditions allow, the water-resistive barrier can also act as the system vapor retarder and air barrier.
- Design and installation of two strong vapor retarders within the wall system is generally avoided. Water that bypasses the water-resistive barrier by way of a leak or condensation becomes entrapped within the system without a means to dry to either side.

The use of a low-permeance membrane in colder climates requires a rigorous analysis of the proposed wall system's hygrothermal (thermal moisture) properties and performance to ensure that moisture will not condense and accumulate within the assembly or become trapped in the wall system without the potential to dry.

Computer Modeling of Temperature and Moisture

For the case histories presented in this paper, we used the WUFI® 4.0 computer program by the Fraunhofer Institute for Building Physics to evaluate the performance of typical wall sections. This program calculates transient (time-varying) thermal and moisture transfer in one-dimensional envelope assemblies. WUFI® uses historical, hourly weather data, including rain, solar radiation, temperature, and relative humidity (RH), to realistically simulate conditions to which the wall will be exposed. We illustrate our modeling work graphically by plotting RH and water content. The dark lines on the graphs represent the profile through the material at the end time for the model. The shaded areas represent the range of conditions that occurred over the entire simulated time period. The light gray shade is a measure of water content and the dark shade is a measure of relative humidity. Surfaces that are indicated to reach 100 % relative humidity during the seasonal cycle are at the dew point temperature of the air within the wall assembly and condensation with surface wetting is likely. Keeping the relative humidity of moisture sensitive materials below 75 to 80 % is a reasonable measure to reduce mold potentials; however, the conditions for mold growth depend, in part, on temperature, RH, and time duration [2]. We consider condensing conditions in the models to occur at the 95 to 100 % range (rather than 100 %) to account for uncertainties of model calculations, material properties, and weather assumptions. We accept that rain typically saturates exterior components, creating 100 % RH conditions, and that moisture may accumulate in portions of the model we normally consider to be outside of the water-resistive barrier.

In modeling a wall system, we are able to vary types and initial conditions of building wall components, and interior and exterior conditions to compare the expected moisture at each wall component for each design. Variables modeled include permeance of water-resistive barrier, thickness of insulation layers, built-in moisture in building materials, cold or warm weather cycle for that geographical location, and use of an interior vapor retarder.

Interior conditions (temperature and relative humidity) vary depending on the residents of the building. For the most part temperature ranges are usually small, but the RH varies depending on the number of people living in the space, fresh-air ventilation rate, and length and frequency of cooking, showers, and other occupant activities. The initial interior conditions for residential buildings in our models are set at 20°C (68°F) ± 1° and 45 % ± 10 % RH. The RH in this case is on the higher, and more conservative, end of what we would expect for typical residential buildings.

The wall assembly is modeled for three years in both typical warm and cold year cycles. The initial moisture content of the building materials is set at 80 % RH to replicate the moisture in new materials. To differentiate between the initial dry out of the building materials and the expected moisture in the wall due to the typical weather and interior conditions, we provide graphs for only the last two years of the three-year cycle (we assume the building materials will dry within the first year).

The WUFI software cannot simulate water leakage through the water-resistive barrier. Therefore, when simulating a leak to evaluate drying potential, we set the initial water content in the exterior sheathing to its 100 % RH value. The model is run through the three-year cycle to record the approximate time for the RH in the exterior sheathing to reach 80 %.

Since the metal lath and adhesive for the drainable EIFS do not have significant thermal or moisture migration resistance, they are not included in the model.

Case Studies

Design considerations for an engineered drainable EIFS will be illustrated by two case studies. The first is a high-rise condominium in southern California and the second is a condominium community consisting of multiple mid-rise and townhouse buildings located in Massachusetts. In both cases, we designed and observed construction of a drainable EIFS cladding to replace an existing barrier EIFS cladding.

Southern California Project

This project is a 22 story high-rise condominium in San Diego, CA. The existing building is concrete frame with infill steel-stud walls clad with a barrier EIFS. The building has several large punched openings with ganged window assemblies comprising casements and fixed lites. Renovation goals included the following:

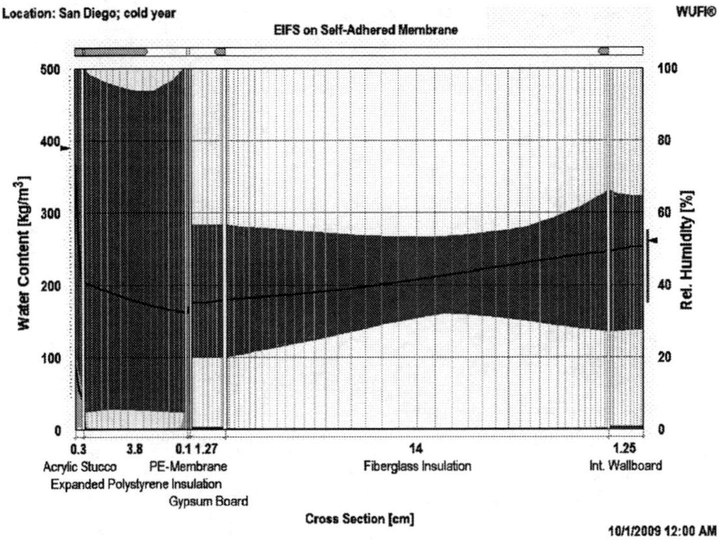

FIG. 3—*Hygrothermal model of EIFS on self-adhered membrane (PE-membrane) in San Diego, CA.*

- Remove mold-contaminated interior gypsum board and finishes and repair or replace corroded steel-stud work.
- Provide new window assemblies as the existing had been field modified (installation of weeps and frame corner seals) previously by others without success.
- Provide a durable and effective water-resistive barrier system including flashings and weeps at each floor level.
- Maintain the existing aesthetics including materials and colors to preclude controversy among so many unit owners.

San Diego, California, is located in a relatively warm mixed climate zone where the temperature does not fluctuate significantly between summer and winter months. The average high temperature during the summer months (June–August) is 23 °C (73 °F) with a high of 37 °C (99 °F). The average low temperature during the winter months (December–February) is 14 °C (57 °F) with a low of 6 °C (43 °F) [3]. The relative humidity and ambient vapor pressure in this coastal region are relatively high throughout most of the year and, as such, we suspected that the wall system has little opportunity to dry to the exterior.

Hygrothermal Analysis—We initially modeled the building wall with the following components (from the interior to exterior): interior gypsum board, 14-cm $\left(5\frac{1}{2}\text{-in.}\right)$ (similar to the existing thickness) glass-fiber batt insulation between steel studs, glass-mat faced exterior sheathing, self-adhered rubberized asphalt/polyethylene membrane or three layers of #15 asphalt-impregnated felt, 3.8-cm $\left(1\frac{1}{2}\text{-in.}\right)$ EPS insulation board and lamina (0.3 cm $\left(\frac{1}{8}\text{ in.}\right)$ of acrylic stucco).

Due to the relatively warm climate and similar vapor pressures on the interior and exterior sides of the wall (i.e., minimal water vapor drive across the wall), a vapor retarder on the inside of the wall system is not necessary. We modeled the system in a cold weather cycle with either a low-permeance membrane (self-adhered rubberized asphalt/polyethylene membrane) or a breathable water-resistive barrier (#15 asphalt-impregnated felt) behind the insulation board with the following results:

1. For the self-adhered rubberized asphalt/polyethylene membrane model (after the initial dry out of the building materials) the RH on the exterior sheathing peaked just above 55 % and the interior gypsum board peaked just above 65 % (Fig. 3). Modeling a leak past the water-resistive barrier, the RH on the exterior sheathing falls to 80 % (from 100 %) in less than six months.
2. For the model with asphalt-impregnated felts under the same conditions as 1, the RH on the exterior sheathing peaked just above 55 % and the interior gypsum board peaked just below 70 %

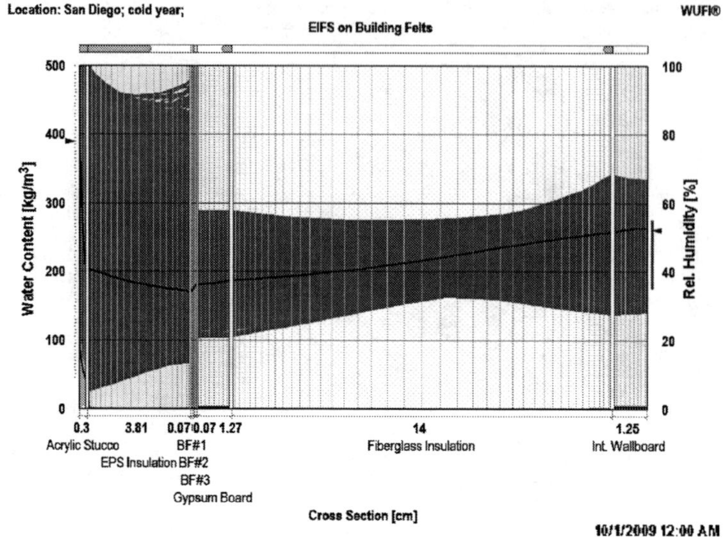

FIG. 4—*Hygrothermal model of an EIFS on three layers of #15 asphalt-impregnated felts (BF #1-3) on exterior gypsum wallboard in San Diego, CA.*

(Fig. 4). Modeling a leak past the water-resistive barrier, the RH on the exterior sheathing falls to 80 % in less than five months.

3. We also modeled the wall system with a polyethylene interior vapor retarder (0.1 perms) and no water-resistive barrier to assess leakage and drying potential to the exterior. In a cold year (where the system is more capable of drying to the outside), the relative humidity at exterior sheathing does not fall below 80 % until twelve months after the initial wetting event (Fig. 5).

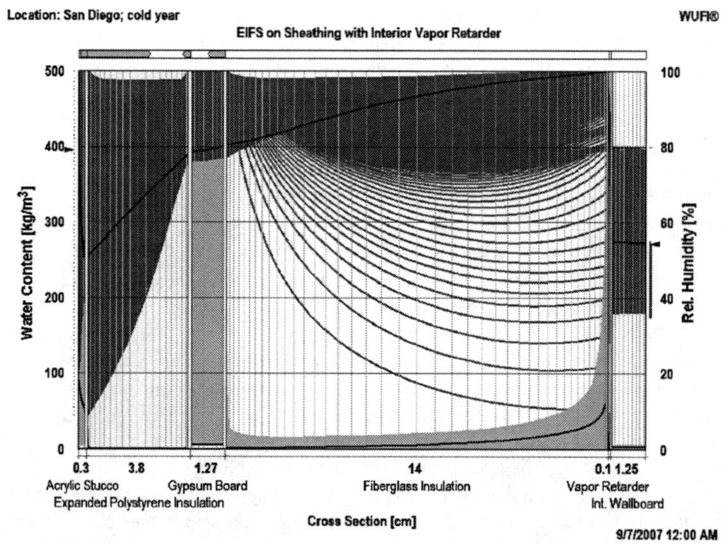

FIG. 5—*Hygrothermal model of an EIFS on wetted exterior sheathing (gypsum board) with an interior vapor retarder in San Diego, CA.*

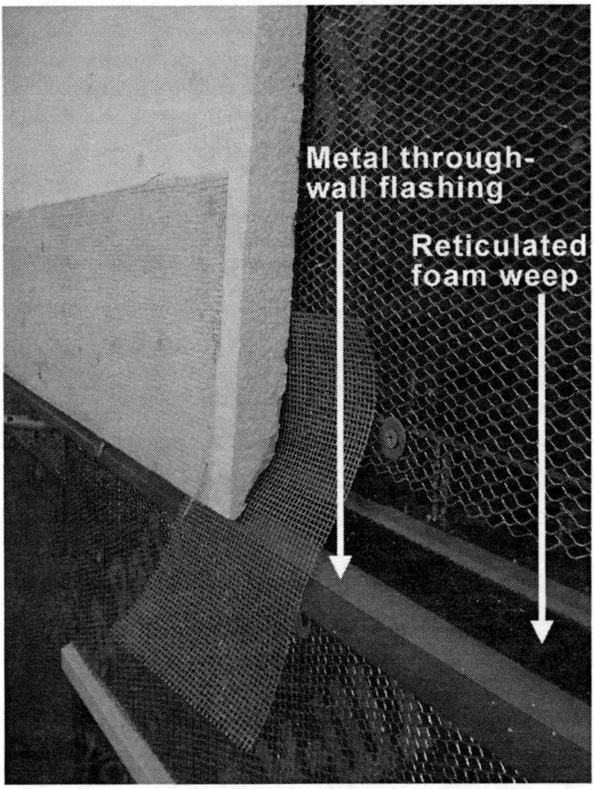

FIG. 6—*Floor line flashing at high-rise condominium in San Diego, CA.*

Both water-resistive barriers keep the RH on the moisture-sensitive substrates below the threshold for possible mold growth. Neither water-resistive barrier type had an advantage over the other in terms of the wall systems' ability to dry out following a leakage event because drying to the exterior is not affected by the permeance of the water-resistive barrier. Due to a lack of moisture drive issues, we chose the self-adhered rubberized asphalt membrane for the water-resistive barrier due to its strong waterproofing characteristics and ease of constructing flashings.

Design Details

Insulation Attachment—Negative wind loading for the building was calculated to be a maximum of 40 psf (1.9 kPa) at the salient corners of upper floor levels. Due to the high wind loads and manufacturer's recommendations, we chose an adhered system to distribute the wind loads over the entire area of the insulation board. Given our selection of a self-adhered rubberized asphalt/polyethylene membrane, we needed to adhere the boards to a mechanically fastened lath, in this case, a G-90 coated steel. This system has the advantage of providing a continuous drainage plane, but requires the use of fasteners that penetrate the self-adhered membrane. The self-sealing nature of the membrane at fastener penetrations eliminates this concern; however, it is imperative that any fasteners for the lath that do not anchor properly be removed and the fastener hole sealed in place. Although the manufacturer's test data indicated that an adhered system had adequate strength to resist the design wind loads, we chose to use supplemental mechanical fasteners with 5-cm (2-in.) wide washers installed through the insulation within 3 m (10 ft) of all salient corners.

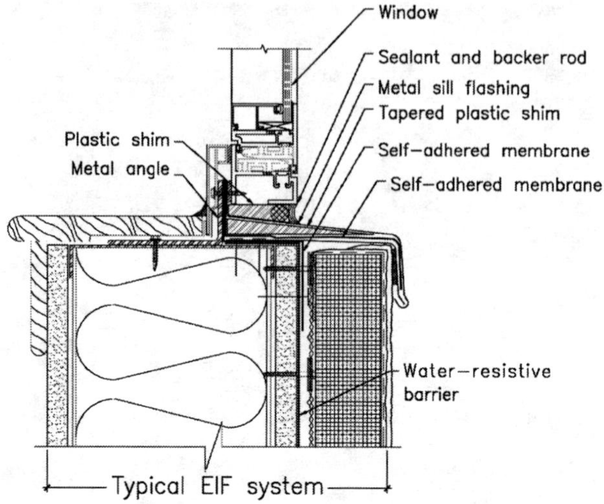

FIG. 7—*Window sill section showing a self-adhered membrane and metal sill flashing.*

Flashing at Floor Lines—This project has horizontal control joints at each floor level to accommodate differential movement between stories. The numerous punched window openings are recessed from the face of the wall with deep returns. As such, we chose the horizontal control joints at floor lines for the flashing/weep location as it simplified the flashing construction and eliminated the additional horizontal line that would result at each floor level if the flashing were installed coincident with the window heads. Figure 6 shows the detail at the horizontal control joint; the insulation board is beveled to help hide the metal flashing, which runs continuously with only the need to splice pieces of metal at the transverse seams and the reticulated foam is installed continuously without a sealant joint to interrupt the flow of water out of the drainage plane. Care must be taken not to clog the drainage plane above the reticulated foam with the insulation board adhesive.

Window Openings—Given our use of a self-adhered rubberized asphalt/polyethylene membrane as the water-resistive barrier, we used the same material to waterproof the window openings. The primary sill flashing was made of membrane that shingled into the drainage plane. Water collected by the flashing is intended to flow in the drainage plane to the floor line flashing weep below. We also designed a simple metal sill flashing that served as a metal sub-sill to limit water infiltration to the backup and to protect the top of the EIFS cladding below the window from run-off and erosion (Fig. 7).

Massachusetts Project

This project is located in a suburban area outside Boston, Massachusetts. It consists of multiple mid-rise buildings and townhouses surrounded by wooded areas. The tallest structures are only three stories; however, the geometric complexity of their exterior facades results in a number of complicated details at the balconies, wood decks, windows, doors, sloped roof systems, mixed cladding systems, and many inside and outside corners. Renovation goals included the following:

- Removal and replacement of rotted plywood sheathing and repair or replacement of rotted wood stud work.
- Interior finishes and the polyethylene vapor retarder were generally to remain in place.
- Remove existing flanged window systems, modify to receive sill flashing, and reinstall windows.
- Provide a durable and effective waterproofing system including flashings to address the many transition details between facade elements.

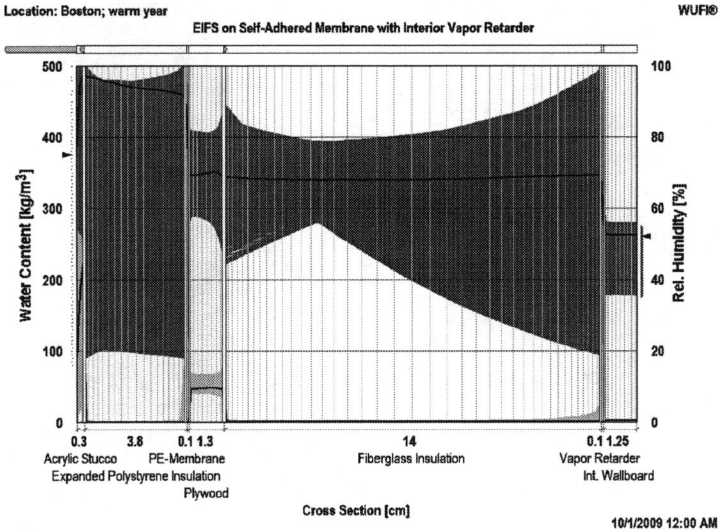

FIG. 8—*Hygrothermal model of an EIFS with a self-adhered membrane (PE-membrane) on plywood with an interior vapor retarder in Boston, MA.*

- Maintain the existing aesthetics, including materials and colors, to preclude controversy among so many unit owners.

During demolition, we observed significant deterioration of plywood sheathing and wood studs from water entrapment within the barrier EIFS, which was originally constructed in the early 1990s. The majority of deteriorated wood was located under windows, kick-outs, decks, and walls shaded from the sun.

Massachusetts is located in a heating climate zone. The coldest winter months extend from the beginning of December to the end of February. The average low temperature from December through February during this most recent winter was −3°C (27°F), and the lowest temperature dropped to −13°C (9°F). The summers are hot and humid and last approximately two to three months, from the end of June to the end of August. The average high temperature for July and August during this past summer was 27.5°C (81.5°F), reaching a high of 36°C (97°F) on some days [4].

The Massachusetts State Building Code sets requirements for water-resistive barriers, insulation, and vapor retarders. The code requires a weather-resistant covering on exterior walls to resist wind and rain. The energy conservation section requires a minimum R-11 in stud cavity insulation for wood-framed walls. The vapor retarder must be installed at the winter warm side of walls and have a maximum permeance of 0.1 perm. This requirement is much stricter than most other codes which specify a permeance rating of 1 perm or less. The water-resistive barrier is to be ten times more permeable than the interior vapor retarder. The interior vapor retarder can be eliminated if the potential condensing surfaces in the envelope system maintain a temperature above the dewpoint temperature of the occupied space.

Hygrothermal Analysis

The existing backup wall system includes the following components (from the interior to exterior): gypsum board finish, polyethylene vapor retarder with 0.1 perm rating, 14-cm (5.5-in.) batt insulation between wood studs, 1.25-cm ($\frac{1}{2}$-in.) plywood. We modeled a similar EIF system as described in the previous case study over the existing wall: self-adhered rubberized asphalt/polyethylene membrane or three layers of #15 asphalt-impregnated felt, 3.8-cm ($1\frac{1}{2}$-in.) EPS insulation board and lamina (0.3 cm (0.12 in.) of acrylic stucco).

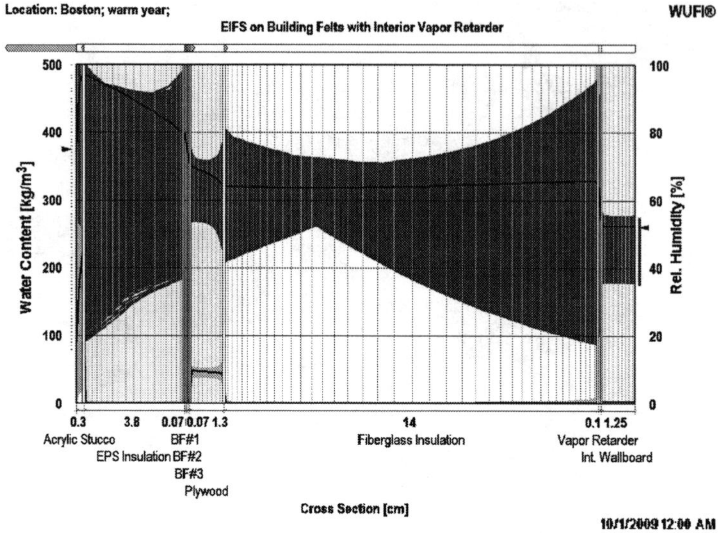

FIG. 9—*Hygrothermal model of an EIFS with three layers of #15 asphalt-impregnated felts (BF#1-3) on plywood and an interior vapor retarder in Boston, MA.*

We modeled the system with either a low-permeability membrane (self-adhered rubberized asphalt/polyethylene membrane) or a breathable water-resistive barrier (three layers of #15 asphalt-impregnated felt) behind the insulation board. We modeled the system in a warm weather cycle with the following results:

1. For the self-adhered rubberized asphalt/polyethylene membrane model (after the initial dry out of new building materials), the RH peaked below 90 % at the exterior plywood and just below 100 % at the interior vapor retarder (Fig. 8). For the same system with a leak through the water-resistive barrier, the RH in both the plywood and interior vapor retarder remain over 100 % after the first year of dry out.
2. For the asphalt-impregnated felts model under the same conditions as in 1, the RH on the exterior sheathing peaked just above 80 % and the interior vapor retarder reached 95 % (Fig. 9). In a leak situation, the RH in both the plywood and interior vapor retarder remain over 100 % after the first year of dry out.
3. For both systems, the RH on the interior vapor retarder reaches the threshold for condensation (95 to 100 %) during the summer months. These warm cycle models illustrate an often overlooked performance problem of using an interior vapor retarder in heating climates with hot humid summers.

We modeled both water-resistive barrier systems without the interior vapor retarder during cold weather cycles (due to increased drying towards the outside) with the following results:

1. For the self-adhered rubberized asphalt/polyethylene membrane model after the initial dry out period, the RH at the exterior plywood peaked just above 90 % and at the interior gypsum board is just above 70 % (Fig. 10). Modeling a leak past the water-resistive barrier, the RH on the exterior sheathing falls to 80 % (from 100 %) in two years.
2. For the asphalt-impregnated felt models under the same conditions as 1, the results are similar to those of the self-adhered membrane models for the typical weather conditions cycle (Fig. 11). The only difference is the RH at the interior gypsum board is just under 70 %. Modeling a leak past the water-resistive barrier, the RH on the exterior sheathing falls to 80 % (from 100 %) in 20 months.
3. The models show that the self-adhered membrane and asphalt-impregnated felt systems performed about equal, with the asphalt-impregnated felts performing better at the interior gypsum board for a leak past the water-resistive barrier. The drying potential for the asphalt-impregnated felt model is better when the vapor drive is towards the outside.

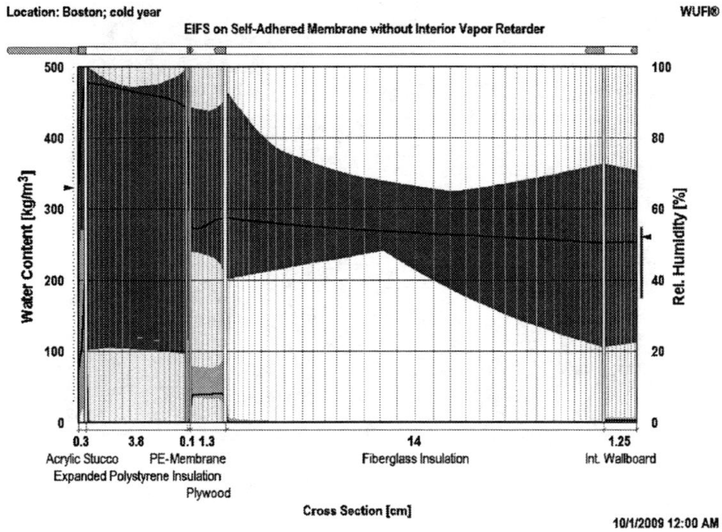

FIG. 10—*Hygrothermal model of an EIFS with self-adhered membrane (PE-membrane) on plywood without an interior vapor retarder in Boston, MA.*

For both membrane systems, the RH on the exterior sheathing remains below condensing levels but in the range that may support mold growth. However, these RH levels occur for approximately two months during the winter when the sheathing temperature is relatively low (0 °C). These RH levels could be reduced by changing the insulation thickness in the stud space to 8.9 cm ($3\frac{1}{2}$ in.), which would make the sheathing warmer in winter while still meeting the overall R-11 value required by code. The results also indicate that the asphalt-impregnated felt system performs somewhat better than the self-adhered mem-

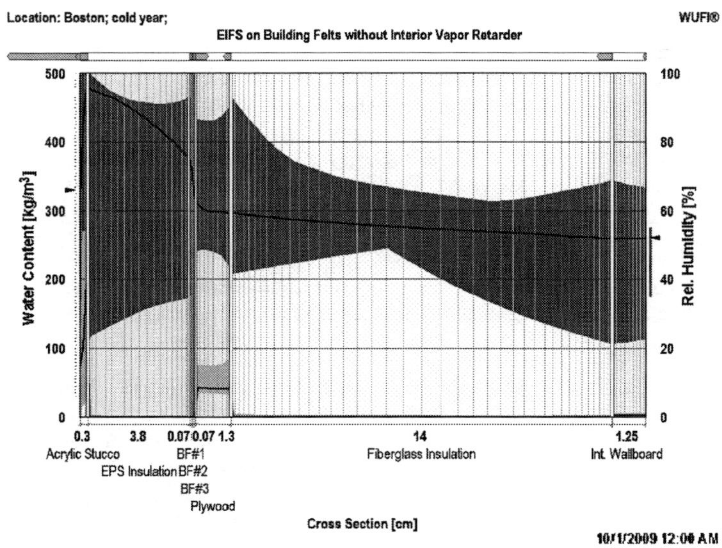

FIG. 11—*Hygrothermal model of an EIFS on three layers of #15 asphalt-impregnated felts on plywood without an interior vapor retarder in Boston, MA.*

FIG. 12—*Self-adhered membrane installed at the window bays and two layers of asphalt-impregnated felts (#15 felt over #30 felt) installed outside the window bays.*

brane because it allows more drying to the outside. However, even in an assembly that has a "breathable" water-resistive barrier and an interior vapor retarder, water between the layers does not readily dry. Overall the wall assemblies without the interior vapor retarder performed best; however, removing the existing interior vapor retarder was financially infeasible for this project. Refer to Ref. [5] for additional models.

For the final wall design, we chose two layers of asphalt-impregnated felt as the water-resistive barrier in the field of the wall, and at window bays, we chose self-adhered membrane shingled from the foundation to the roof line (Fig. 12). This combination provides a breathable barrier at most of the wall surface (avoiding the use of two vapor retarders within the assembly) and a more robust membrane in the areas of highest water exposure.

Design Details

Window Openings—The existing windows, which have a perimeter nailing flange, were reused in an effort to reduce costs. Our water tests showed that the window sill frame corners leak readily, and therefore a functional sill flashing is imperative to capture and drain this water. The flanged windows restricted the use of a traditional metal sill flashing that drains water directly to the outside of the cladding. In order to drain the sill pocket, the window sill nailing flange was field modified to include weep slots to allow water to flow into the drainage plane below the window. The sill flashing is comprised of a self-adhered membrane over a liquid-applied membrane and sloped shim, with back and end dams formed by the membrane adhered onto the wood framing (Fig. 13).

Flashing at Floor Lines (Weeps)—Given the low-rise nature of the facade, we elected to simplify the design by providing a continuous flashing/weep at the first-floor level and at the heads of each window opening.

Complex Transition Details—As is often the case in low-rise residential construction, a number of complex flashing transitions occur at the intersection of various wall components. Figure 14 shows the intersection of a balcony railing wall with the building wall and balcony waterproofing membrane. The wall water-resistive barrier is continuous behind the EIFS on the building wall and cedar siding on the railing wall, and with the use of a metal separation layer, the wall water-resistive barrier counter flashes the balcony waterproofing membrane. With the ability to design and install backup flashings of metal and self-adhered membrane, these transitions can be made watertight and relatively maintenance free. In a barrier EIFS, this type of detail would ultimately rely on a surface sealant joint with a difficult installation geometry, resulting in considerable risk for leakage and ongoing maintenance and repair.

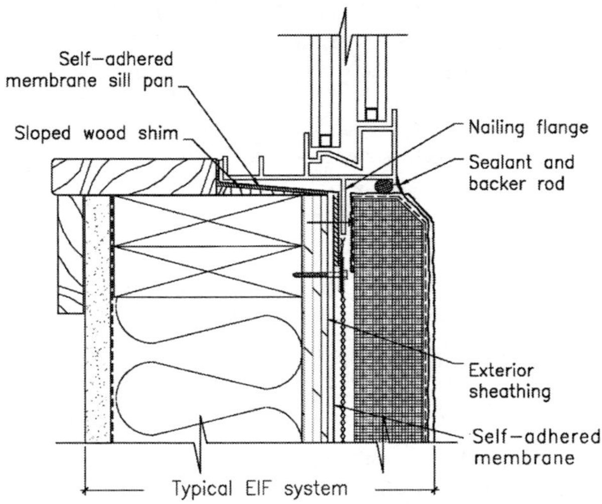

FIG. 13—*Self-adhered membrane sill flashing at a window with a nailing flange.*

FIG. 14—*Self-adhered membrane is installed behind the EIFS at critical transitions, such as the transition from the building wall to the railing wall, and laps over metal flashings.*

Conclusions

An engineered drainage plane EIFS that is properly designed and installed can provide a cladding system with effective and durable waterproofing components with the same aesthetic features as a traditional EIFS that has appealed to many designers and consumers. They are ideally suited as a replacement system for a barrier EIFS where the building structure requires a lightweight cladding and a change in building aesthetics is unacceptable or controversial to the owner.

While manufacturers offer a number of "off-the-shelf" drainable systems that are a significant improvement to a barrier EIFS, some of the water-resistive barriers do not have a proven track record, and some drainage plane designs do not facilitate the flow of water within the assembly. Designers can engineer systems to meet project specific requirements by incorporating waterproofing membranes and drainage grids or laths from a variety of specialty manufacturers into the basic EIFS cladding.

A drainable EIFS gives designers increased flexibility in wall design because the EPS insulation provides an insulating layer out-board of the waterproofing membrane. By adjusting the thickness (R-value) of the EPS insulation and batt insulation, it is often possible to maintain temperatures in all parts of the backup wall above the dewpoint of the interior air during winter. If conditions allow (based upon proper hygrothermal analysis), the waterproofing membrane can also act as the system vapor retarder and air barrier.

References

[1] Ruggiero, S. S. and Bronski, M. B., "Exterior Insulation and Finish Systems (EIFS) Use in Wood-Framed Residential Construction: Design Concepts to Avoid Common Moisture Intrusion Problems," *J. Test. Eval.*, Vol. 28, No. 4, 2000, pp. 290–300.

[2] Sedlbauer, K., "Prediction of Mould Growth by Hygrothermal Calculation," *Journal of Thermal Envelope and Building Science*, Vol. 25, No. 4, 2002, pp. 321–336.

[3] U.S. Department of Commerce, National Oceanic and Atmospheric Administration, National Weather Service, Climate Data for San Diego, CA, USA, September 2005 to August 2006, URL: http://www.weather.gov/climate/getclimate.php?wfo=sgx, National Weather Service Climate Data, 14 September 2006.

[4] U.S. Department of Commerce, National Oceanic and Atmospheric Administration, National Weather Service, "Climate Data for Boston, MA, USA, September 2005 to August 2006," URL: http://www.weather.gov/climate/getclimate.php?wfo=box, National Weather Service Climate Data, 14 September 2006.

[5] Truong, C. (Ching Ching), Cooperative Study Term Paper "An Analysis of Vapour Retarders in Exterior Insulation Finish Systems," University of Waterloo (unpublished), 2006.

Journal of ASTM International, Vol. 5, No. 9
Paper ID JAI100827
Available online at www.astm.org

Linda McGowan,[1] *Scott Riley,*[2] *Jim Connolly,*[3] *and William Shalkowski*[4]

Evaluation of Debonding of Lamina from Foam and Repairs to Exterior Insulation and Finish System Façade

ABSTRACT: A large building in Denver, Colorado was constructed in 1999 with exterior insulation and finish system (EIFS) cladding. About a year after completion, the lamina started to debond from the foam on the darker-colored EIFS panels. Tests were conducted to determine the cause of the delamination. The influence of solar radiation was studied. Although elevated temperatures were measured in these areas, they were not above the "melting point" of the foam insulation. During the investigation it was discovered that the pattern of the fiberglass mesh appeared to be "embossed" in the foam surface. Microscopic examination confirmed that the surface of the foam had been locally dissolved in the pattern of the mesh. Chemical analysis and exposure tests performed on the fiberglass mesh used in the lamina revealed that plasticizers migrating from the mesh would attack the interface surface of the foam insulation at elevated temperatures that are below the typical melting point of the foam. The elevated surface temperatures of the darker-colored panels appeared to accelerate the effect of the plasticizers. The scope of repairs at the darker-colored EIFS included removal of the original lamina, preparation of the foam, and installation of new lamina. The mesh used in the new lamina was analyzed to ensure that it did not contain plasticizers capable of dissolving the foam. The lighter-colored EIFS had not exhibited widespread delamination of the lamina. However, the risk of migration of the plasticizers and the potential for future delamination still existed. Mechanical-fastening was used to secure the original lamina to the framing, then a new lamina was applied over the original lamina. This repair concept was dependent on proper bond between the new and original laminas. The bond was tested using techniques similar to those used for measuring adhesion for both new synthetic and cementitious base coat materials.

KEYWORDS: delamination, E06.58, exterior insulation and finish system, EIFS, lamina, mesh, plasticizer, repairs

Background

The building was originally constructed in the late 1990s; the exterior insulation and finish system (EIFS) installation began in about November 1998 and extended through July 1999. The building is generally oval-shaped in plan, and has exterior facade elements of precast concrete and glass curtain walls, in addition to the EIFS. The long axis of the oval is generally in the east-to-west orientation. The EIFS cladding encompassed approximately 8360 m^2 (90 000 square feet).

The EIFS system is a barrier-type facade system [1] that consists of the following components (from inside the wall to the outside of the wall):

- Steel stud framing supported by a structural steel and cast-in-place concrete framing system.
- Exterior glass fiber-faced gypsum wall sheathing.
- ASTM Type I closed mold expanded polystyrene (EPS) foam insulation [2,3] adhered to the exterior gypsum wall sheathing with an adhesive bonding material 15.24 cm (nominal 6 in.) EPS thickness at the darker-colored EIFS areas and 5.08 cm (2 in.) EPS thickness at the lighter-colored EIFS areas).
- Polymer-based cementitious base coat (PB system), consisting of a blend of portland cement, sand, and a polymer emulsion admixture, into which an intermediate weight (approximately 339 g/m^2 (10 oz/sy) glass-fiber mesh is embedded, with a total thickness of approximately 2.54 cm (1/16 in.).

Manuscript received October 11, 2006; accepted for publication September 15, 2008; published online October 2008.
[1] Building Consultants & Engineers, Inc., Littleton, CO.
[2] Wiss, Janney, Elstner Associates, Inc., Lakewood, CO.
[3] Wiss, Janney, Elstner Associates, Inc., Northbrook, IL.
[4] Big Horn Plastering, Inc., Littleton, CO.

FIG. 1—*Debonding of lamina from foam.*

- Polymer finish coat, consisting of pigment, a polymer emulsion, and aggregate. All EIFS products were supplied by the EIFS manufacturer.

The EIFS had three colors: rust, dark beige, and light beige. The rust-colored EIFS had relatively deep reveals in the foam insulation 7.62 cm (about 3 in.) deep by 5.08 cm (2 in. wide) and created a "panelized" appearance for much of the façade. The dark and light beige-colored EIFS had relatively shallow V-notch reveals for a similar effect. Glass-fiber "detail" mesh weighing approximately 145 g/m^2 (10 oz/sy) was used to reinforce the lamina in the reveals.

It had been reported that debonding of the lamina was first visually observed in about September 2000. The areas of debonding were located primarily in the rust colored EIFS, which is generally located on the east, south, and west elevations. Figures 1 and 2 illustrate the debonding of the lamina that was visually observed, which could be described as "sagging" of the lamina.

Initial Investigations

An initial investigation of the debonding lamina began in April 2001 by an independent engineering/architectural investigation firm. The report on the initial investigation claimed visual evidence that portions

FIG. 2—*Debonding of lamina from foam.*

of the EIFS lamina had debonded from the EPS and was most common in portions of the rust-colored cladding located on the east, south, and west elevations. The report stated that EIFS areas oriented toward the north elevation did not appear to be debonding.

Among other activities, the initial investigation included a visual survey to identify problem areas and the retrieval of 30 samples from the building. Seventy tensile bond evaluations (pull tests) were performed using a dynamometer to evaluate the bond and degree of adhesion between the EIFS components, as well as the failure mechanism of the lamina.

The initial evaluation concluded that the EPS board had softened or melted at various locations and that the failure of the lamina was primarily the result of "melting" of the EPS board as the result of elevated temperatures caused by solar heat gain of the dark-colored EIFS. Repair recommendations included removal of the EIFS lamina and insulation board for much of the south, east, and west-facing elevations and replacement with another non-EIFS alternative. The recommendations did not require any modification to the existing EIFS on the north elevation.

Subsequent investigations began in 2003 by a second independent engineering/architectural investigation firm. The subsequent investigations included visual examination of the building and the retrieval of a number of EIFS samples, measurement of base coat thickness, "pull tests" at various locations, and temperature monitoring of the building at a number of locations. This included monitoring of ambient air temperatures, as well as thermocouples to measure the temperature of the EIFS at the lamina-to-foam interface. Temperature monitoring was performed between August 20 and September 25, 2003. Maximum recorded temperatures of the EIFS lamina-to-foam interface were 86°C (186.8°F) at a southwest-facing location adjacent to a reflective surface and 83.83°C (182.9°F) at a west-facing location not exposed to reflective light. Laboratory evaluations were also performed, including placing mesh samples in direct contact with EPS board and exposing the system to elevated temperatures and exposing a number of mock-up EIFS panels to elevated temperatures.

The subsequent investigation concluded that the temperature of the EPS was not the primary cause of the delamination of the lamina, although it was accelerated by the temperature conditions in the dark red areas. The evaluation concluded that the delamination was due to migration of a component in the lamina, which etched and degraded the EPS board surface. Specifically, the investigation points to a plasticizer used on the mesh in the original lamina. The report noted that for the light beige-colored areas delamination caused by plasticizer migration could still occur. Repair recommendations included complete removal and replacement of the lamina in the rust-colored areas, the overlay of 2.54 cm (1 in.) thick EPS foam, mechanically attached to the structure through the existing EIFS and covered with new lamina at the dark beige and light beige areas.

The findings of the subsequent investigations indicate that ambient temperature does not necessarily correlate with maximum temperatures of the base coat; that solar radiation likely has a greater impact on the temperature of the base coat for darker EIFS colors than does ambient temperature; and that radiant heat from reflective surfaces (i.e., roof top units, glazing, etc.) further increases the temperature of the base coat.

Laboratory Testing

Based on a review of the conflicting information from the initial and subsequent investigations, the authors of this paper were charged with developing opinions as to the cause of the delamination and establishing a repair scope and procedure. As part of this evaluation process, laboratory testing was undertaken, which focused on the plasticizer of the original mesh [4,5].

Scope of Testing

Testing procedures were developed to determine several important characteristics of the original mesh used on the building. These tests were intended to determine the following: the chemical composition of the plasticizer used on the mesh, if and at which temperature the plasticizer deteriorates the foam insulation, and if the plasticizer deteriorates the foam insulation without direct contact or the addition of heat. Following the completion of these tests, additional testing was performed to determine the temperature at which the foam insulation deteriorates and if the chemical composition of the plasticizer used in the

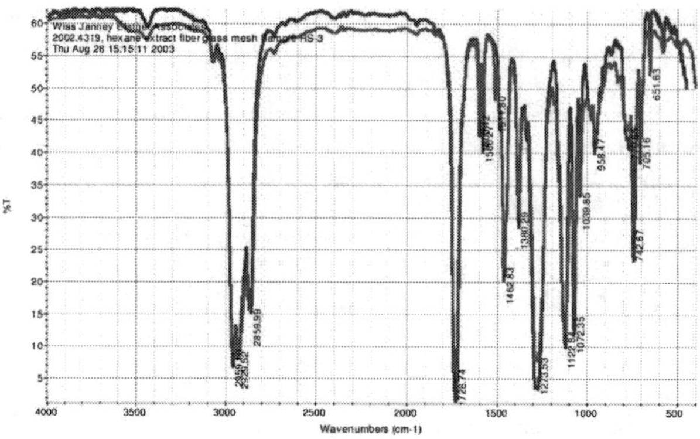

FIG. 3—*FTIR spectrum of hexane extract of mesh from Sample RS-3 (red) compared to reference spectrum of dioctyl phthalate (maroon).*

coating on the mesh differed from that used in the coating on similar mesh currently supplied by the same EIFS manufacturer, as well as that used on similar mesh supplied by other leading EIFS manufacturers.

The testing procedures utilized are outlined as follows:

Part 1: To determine the chemical composition of the plasticizer used on the mesh:

- Prepare mesh sample from actual sample taken from the building.
- Extraction to separate plasticizer from sample.
- Fourier transform infrared spectroscopy (FTIR) analysis to identify the plasticizer.

Part 2: To determine if and at what temperature the plasticizer deteriorates the foam insulation:

- Plasticizer etching experiments at 24°C (75°F) and 37.78°C (100°F).
- Photomicrography.

Part 3: To determine if the plasticizer deteriorates the foam insulation without direct contact:

- Exposure to plasticizer vapor at 60.00°C (140°F)

Part 4: To determine if the chemical composition of the plasticizer used in the coating on the mesh from an actual sample taken from the building differs from that used in the coating on similar mesh currently supplied by the same EIFS manufacturer, as well as that used on similar meshes by other leading EIFS manufacturers:

- Prepare each mesh sample.
- Extraction to separate plasticizer from each mesh sample.
- FTIR analysis to identify the plasticizer from each sample.

Part 5: To determine at what temperature the foam insulation deteriorates:

- Exposure of foam insulation in oven at 60.00°C (140°F), 71.11°C (160°F), 82.22°C (180°F), 87.78°C (190°F), and 93.33°C (200°F).

Testing Results

Part 1: Extraction and Infrared Spectroscopical Analysis—The glass-fiber mesh was removed from one of the samples taken from the building, and the adhering cementitious base coat material was separated from the mesh. The mesh was then extracted with hexane to separate any plasticizer from the coating on the mesh. The extract was dried at low temperature, and then analyzed using Fourier transform infrared spectroscopy (FTIR). Based on the infrared spectrum of this extract shown in red in Fig. 3, the plasticizer was identified as a phthalate ester similar to dioctyl phthalate (shown in maroon).

Part 2: Compatibility Studies—Droplets of the plasticizer (oil) were applied to the surface of the polystyrene foam insulation from a sample that had been removed from the building. After 16.5 hours of exposure at ambient temperature 23.89°C (75°F), the surface of the board was examined visually and microscopically. The areas where the plasticizer had been applied were slightly depressed due to the reaction of the plasticizer with the foam insulation, as shown in the photomicrograph in Fig. 4, which was

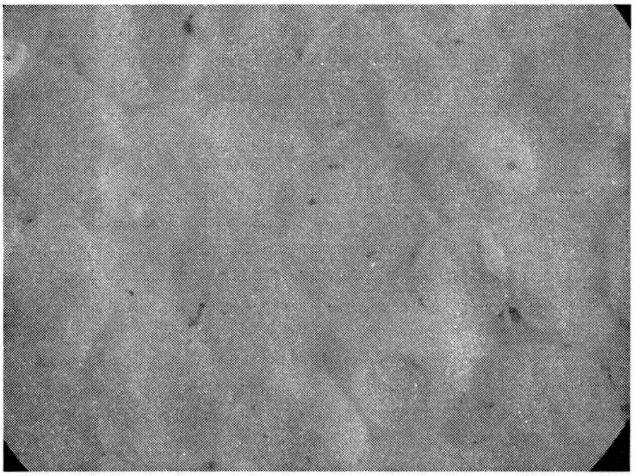

FIG. 4—*Photomicrograph of slightly depressed polystyrene foam insulation where plasticizer was placed, 16.5 hours at 23.89°C (75°F).*

taken at 10× magnification. The foam insulation sample was then placed in an oven at 37.78°C (100°F) for approximately 6.25 hours. After this exposure, the sample was re-examined, and the areas where the plasticizer had been applied were dramatically depressed (approximately 2.54 cm (1/16 in.) to 2.54 cm (1/8 in.) as indicated in the photomicrograph in Fig. 5, also at 10× magnification.

Part 3: Evaluation to Determine If the Plasticizer Deteriorates the Foam Insulation Without Direct Contact—The plasticizer oil extracted from a mesh sample taken from the building was placed in a 50 mL beaker and a section of polystyrene foam was placed on top of the beaker. This assembly was placed in an oven maintained at 60.00°C (140°F) for one week to determine if any volatile components (vapor) from the plasticizer would attack the polystyrene board. No evidence of chemical attack was detected visually after one week of exposure. Thus, direct contact between the plasticizer and the polystyrene foam board had a much more dramatic effect on the foam board (i.e., relatively deep etching of the foam at only 37.78°C (100°F) after 6.25 hours).

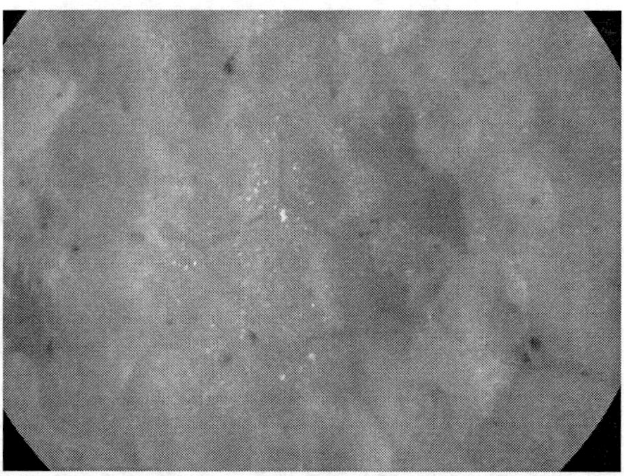

FIG. 5—*Photomicrograph of greater depression in polystyrene foam insulation where plasticizer was placed, 6.25 hours at 37.78°C (100°F).*

TABLE 1—*Comparison of extraction residue from mesh samples.*

Mesh Brand	Extraction Residue (plasticizer) %/wt. of mesh	Identification
Original Brand (Same as used in original construction)	3.50	Phthalate ester (similar to dioctyl phthalate)
Brand A	0.19	Crystalline wax with a small amount of a natural oil
Brand B	<0.10	Silicone-modified hydrocarbon
Brand C	0.11	Styrene butadiene and surfactant

Part 4: Comparative Study of Chemical Composition of Plasticizers—A study was undertaken to determine if the mesh supplied by other leading EIFS manufacturers contained similar plasticizers. Accordingly, samples of unused, intermediate (medium) weight, fiberglass mesh supplied by the other leading EIFS manufacturers were obtained and submitted for analysis.

These samples were processed and extracted with hexane to separate any plasticizing oils from the coatings on the mesh. Those extracts were then dried at low temperature and analyzed using Fourier transform infrared spectroscopy (FTIR).

The types and amounts of extracted material are given in Table 1.

Part 5: Exposure of EPS to Elevated Temperatures in Oven—A portion of an expanded polystyrene foam insulation sample taken from the building was cut into sections and exposed to temperatures ranging from 60.00°C (140°F) to 93.33°C (200°F). A section was retained at room temperature for comparison purposes. Figures 6 and 7 show the results, and the observations made are listed in Table 2.

The testing of the foam insulation was performed to address the issue of heat build-up and potential "melting" of the foam from heat only. Maximum temperatures measured during the investigation by others were less than 87.78°C (190°F), although some data received at a later date reflected the maximum temperature in one localized area may have been as high as 91.67°C (197°F) for a short period. The foam was tested to various temperatures for a number of days. No observable damage was noted for the foam at

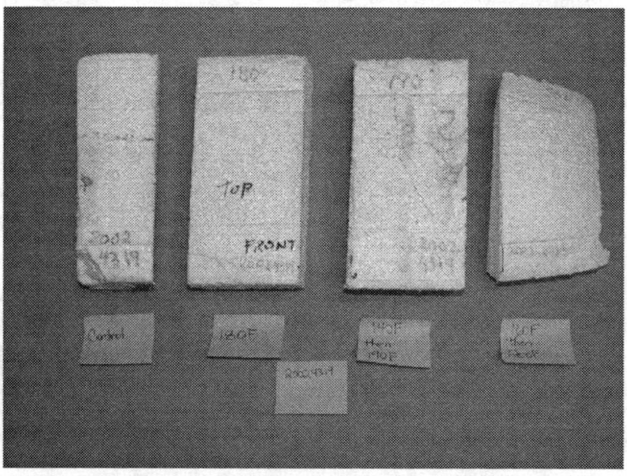

FIG. 6—*Front view of foam insulation after testing.*

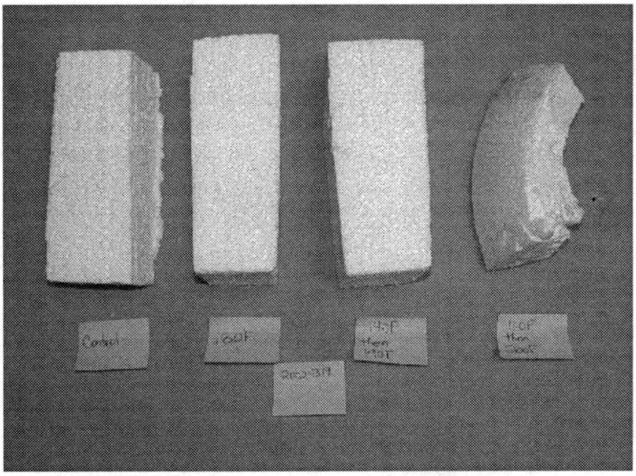

FIG. 7—*Side view of foam insulation after testing.*

87.78°C (190°F) or less; damage was observed for foam exposed to a temperature of 93.33°C (200°F).

For reference, the foam insulation was specified to conform to ASTM C578 Type IV, [1] which is indicated to maintain dimensional stability at a temperature of 73.89°C (165°F).

Conclusions

The laboratory tests performed indicate that the plasticizer in the original mesh occurred in a much higher quantity than in similar mesh provided by other leading EIFS manufacturers. In addition, the testing demonstrated that the plasticizer in the original mesh deteriorated the EPS board at relatively low temperatures, and the rate of deterioration increases at elevated temperatures. Due to the incompatibility between the foam and the plasticizer, the foam would deteriorate at relatively low temperatures (lower than the rated temperature for this type of EPS board). The mesh, or more specifically the plasticizer used on the mesh, therefore was determined to be the primary catalyst in the deterioration of the EPS board and the debonding of the lamina.

The laboratory tests, in conjunction with the observations and temperature monitoring performed by others, did not support the hypothesis that melting of the foam and resultant debonding of the lamina was caused only by heat.

TABLE 2—*Results of exposure of EPS to elevated temperatures.*

Temperature, (F)	Number of Days	Observations
60.00°C (140)	23	No visible changes. Moved to 87.78°C (190 °F) oven.
71.11°C (160)	23	No visible changes. Moved to 93.33°C (200 °F) oven.
82.22°C (180)	23	No visible changes.
87.78°C (190)	23@ 60.00°C (140 °F) Then 7@ 87.78°C (190 °F)	No visible changes.
93.33°C (200)	23@ 71.11°C (160 °F) Then 7@ 93.33°C (200°F)	Foam bowing and shrinking internally after 24 hours at 93.33°C (200°F), possible slight continued shrinkage Day 2 to Day 7.

Solar heat gain played a significant part in higher base coat temperatures, as did the darker EIFS colors. These elevated temperatures accelerated the migration of the plasticizer and increased the rate and amount of deterioration of the EPS and, therefore, contributed to the debonding of the lamina. Reflective surfaces adjacent to the darker-colored EIFS also contributed to higher base coat temperatures, which increase this phenomenon.

A number of installation factors, such as inadequate basecoat thickness, improperly prepared EPS substrate, insufficient lap area at splices, and gaps in the insulation board were likely secondary contributory factors for the failure of the EIFS in certain of the samples that were taken.

Repair Recommendations

The initial repair recommendation included the complete removal of the existing lamina on all areas of the original EIFS and the installation of new lamina with a mesh that does not have the detrimental plasticizer. This recommendation stems from the fact that mesh with the detrimental plasticizer was used in all areas of the building.

In order to evaluate alternative repair options, consideration was given to installing a new layer of 2.54 cm (1 in. thick) EPS foam mechanically attached to the exterior of the building at the dark beige and light beige areas, covered by a new lamina, as recommended during the subsequent investigation by others. This was considered to be a viable option, if it could be successfully accomplished, and no long-term problems would be associated by leaving the original lamina and mesh in place.

Cost Estimate

A cost estimate was developed for the recommended repair work by an independent cost estimator. Two schemes were evaluated: (1) complete removal and replacement, and (2) removal and replacement of the lamina on the darker-colored portions of the building that faced south, east, and west and installation of a new lamina over a new layer of 2.54 cm (1 in.) thick EPS mechanically attached through the existing EIFS on the lighter-colored, northern portion of the building. The initial cost estimate indicated that the costs were very close (within 5 %) for these two repair schemes, and both had the same approximate estimated time frame to complete the work.

Because of a number of aesthetic issues (such as possible curvature problems with the new 2.54 cm (1 in.) thick EPS on the slightly curved exterior walls) and installation issues (such as the length of the screws, flashing details, and joint/reveal details) associated with the proposed overcladding approach, the EIFS repair contractor suggested installing a new lamina directly over the existing lamina on the lighter-colored, northern portion of the building. After evaluating this scheme, it was decided to omit the additional 2.54 cm (1 in.) thick layer of foam insulation and apply the new lamina over the existing lamina on the north side of the building. This resulted in a simplified assembly, a faster repair procedure, and a significant cost savings.

Development of Repair Documents

Removal and Replacement of Lamina

Following the decision to replace the lamina on a large portion of the building, specifications were developed indicating the materials and general repair procedures to be used. This repair method was utilized at the rust and dark beige-colored EIFS that had the 15.24 cm (6 in.) thick EPS foam; and at other areas of debonded lamina, regardless of color or foam thickness. These conditions typically occurred on the east, south, and west elevations. The specifications outlined how the existing lamina was to be removed, and how the existing foam insulation that was to remain should be prepared to receive the new lamina materials.

The repair specifications called for the new lamina materials to be provided by a different EIFS manufacturer than those used in the original construction. The new mesh materials did not contain the same plasticizer on the mesh used in the original construction. The new lamina system consisted of an intermediate weight mesh (407 g/m^2 or 12 oz/sy) in a base coat followed by a standard weight mesh

(146 g/m^2 or 4.3 oz/sy) in another base coat applied over the prepared foam insulation. Detail mesh (146 g/m^2 or 4.3 oz/sy) was to be used at the reveals and joints. The base coat for the new lamina was a polymer-modified cementitious base coat, and the finish coat was a light pebble-aggregate acrylic-based finish coat. The color of the new finish coat was specified to match the original colors for the building (not the actual current colors on the building, since fading had occurred). To minimize fading of the new finish coat a water-based acrylic color coating was specified at the rust and dark beige-colored lamina. It should be recognized that the manufacturer of the new EIFS materials was involved during the development of the design and their input regarding the selection of materials, application methods, and procedures, was useful given the unique nature of the repairs for this building.

The existing sealants were specified to be completely removed and replaced. New primer, backer rod, and silicone sealant was specified for use in the joints. The color of the new sealant was specified to match the original colors for the building. Because of the EIFS materials and sealant that were used, it was possible to achieve a ten-year materials and labor warranty for the work. This was provided by the EIFS manufacturer.

Installation of New Lamina over Existing Lamina

It was determined that new lamina could reasonably be installed over the existing lamina at the light-colored EIFS, which had been installed over 5.08 cm (2 in.) thick EPS foam and was located primarily on the north elevation. In these areas, materials and methods were evaluated to determine which would be the most effective at cleaning the surface of the existing lamina without adversely affecting the mechanical fasteners. The intention was to clean the existing lamina to sufficiently remove dirt and other contaminants, without harming the existing lamina or zinc-coated steel fasteners. It was determined that a mild solution of trisodium phosphate (TSP) was reasonably effective. A proprietary EIFS cleaner was also considered, but ultimately rejected, since some of its components had the potential to be detrimental to the fasteners.

Cleaning of the existing lamina required that the surface be prewetted prior to the application of the cleanser. The cleanser was then applied and lightly scrubbed into the lamina surface. It was important for the cleanser to be completely rinsed from the surface of the existing lamina. Since water could otherwise enter into the EIFS system, the existing sealants could not be removed prior to cleaning the lamina. In addition, the existing lamina was required to be completely dry after cleaning prior to applying the new lamina.

Because this repair approach was dependent on the adhesion between the new and existing lamina, there was concern regarding the bond strength of the new base coat materials to the existing lamina that was to remain. Testing was undertaken to ensure the new base coat would adhere to the existing lamina [6]. Two types of base coat materials were evaluated: polymer-modified cementitious base coat and polymer-based noncementitious base coat. Panels with each type of base coat with standard weight mesh were constructed on the cleaned and dry existing lamina.

Testing was conducted with an elcometer adhesion tester. The test process consisted of isolating a 7.62 cm (3/4 in.) diameter sample using a hole saw to cut around its perimeter, carefully cutting through the new lamina and existing finish coat, but not cutting into the existing base coat and mesh. Then, a dolly was epoxied onto the sample area. The testing apparatus was carefully mounted onto the wall to engage the dolly, and the "pull" of the lamina was measured. The failure mechanism (between lamina layers or between the lamina and the foam) was noted. Lamina "pull" values were noted. It was concluded that the noncementitious base coat adhered better to the existing lamina than the cementitious base coat, and the bond between the new and existing lamina was sufficient. Therefore, the base coat for the new lamina was specified to be a polymer-based noncementitious base coat.

Fastener spacing was based on a grid utilizing the spacing of the existing steel studs and the new lamina manufacturer's recommendations. The fasteners were specified to be zinc-coated self-drilling steel fasteners with plastic EIFS attachment washers. Since this repair method was only used where the existing foam thickness was 5.08 cm (2 in.), the fasteners were about 10.16 cm (4 in.) long.

It is important to note that for a portion of the building this repair technique was successfully employed in an overhead application.

FIG. 8—*Removal of the lamina in the reveals resulted in significant damage to the foam along the shoulders of the reveal.*

Repairs to Building

Removal and Replacement of Lamina

Prior to full scale execution of the repair work, mock-up areas were selected to evaluate the specified repair approach. During this process, it was discovered that despite significant delamination of the lamina within the field of the EIFS, the lamina within the reveals remained well bonded to the underlying foam. As shown in Fig. 8, removal of the lamina in the reveals resulted in significant damage to the foam along the shoulders of the reveal, which would require rebuilding the majority of the reveal profiles. Based on the condition of the lamina at the reveals and the impact this would have had on the repair schedule and cost, the mock-up procedure was modified.

In the revised repair approach, the lamina was saw cut approximately 2.54 cm (1 in.) away from the edge of the reveal. The existing lamina was cut to a depth equal to, but not greater than, the existing lamina thickness. The field of the lamina was then removed by "peeling" the lamina away from the foam insulation. The lamina that remained in the reveal was examined to verify that all edges remained well bonded to the foam. Where the lamina adjacent to the reveal was not well bonded, that section of the lamina was cut out and the lamina and foam repaired. At the completion of this process, the lamina had been removed from the field of each panel exposing the original foam insulation, and the panels were bordered by a "picture frame" of the original and, in limited locations, repaired lamina that remained in the reveals. The removal of the lamina resulting in a "picture frame" appearance is shown in Fig. 9.

Within the field, any loose or damaged foam was replaced, gaps were filled with foam slivers, and the entire surface of the foam was rasped in preparation for installation of the new lamina. Cementitious base coat was installed over the prepared foam, and intermediate weight mesh was troweled into the base coat up to the edges of the existing lamina that remained along the reveals. A second application of cementitious base coat was installed to fully encapsulate the mesh, to build up the thickness of the new base coat flush with the existing lamina that surrounded each panel, and because the mesh pattern bleed through would not allow for only one layer of base coat. The installation of the base coat is shown in Fig. 10.

Cementitious base coat was then extended into the reveals and detail mesh was installed in the reveals and onto the surrounding base coat on either side of the reveals. Once the detail mesh was fully encapsulated, an additional layer of cementitious base coat was applied over the panel field; and standard weight mesh was troweled into this final base coat layer and lapped over the detail mesh. This is shown in Fig. 11.

The existing lamina appeared to be well bonded to the EPS foam within the joints, and there did not appear to be any indication that adhesion at the backwrap details was compromised. Therefore, the existing

FIG. 9—*Removal of the lamina resulted in a "picture frame" appearance.*

lamina in the joints was allowed to remain so that the foam did not need to be removed and replaced to accommodate backwrapping of the new mesh. There was a significant cost savings associated with allowing the backwrap detail to remain, which otherwise would have required the removal of the existing EPS foam on each side of the joint. The existing sealant was cut out of the joints, and the remaining sealant on the sides of the joint was cleaned off the surface of the existing lamina with a combination of grinder, scraper, and wire wheel. New cementitious base coat and detail mesh extended into the sides of the joint, and were lapped with the base coat and mesh in the panel field. Once the base coat was cured, the sides of the joint were primed, and backer rod and sealant were installed [7]. After the sealant were sufficiently cured, the finish coat was installed. The sealants were tested at various locations to verify that they were properly installed. Pull tests were performed to evaluate sealant thickness and profile. Nonconforming work was corrected.

Installation of New Lamina over Existing Lamina

The installation of the new lamina over the existing lamina first required locating the existing steel studs. This was accomplished by drilling into the wall assembly until a stud was encountered. Once a stud was

FIG. 10—*Base coat application inside the "picture frame."*

FIG. 11—*Base coat and mesh installed over entire wall surface.*

located, vertical chalk lines were placed at 40.64 cm (16 in.) on center to correspond with the stud spacing. Horizontal chalk lines were placed at 30.48 cm (12 in.) on center within the field of the panel, and then the fasteners were installed at the intersection of the chalk lines. When fastening through a metal stud, the screw had sufficient bite to pull the plastic EIFS attachment washer into the lamina which created a slight depression. When this did not occur, it indicated that the stud had been missed; and the line of fasteners were shifted to match the stud spacing. Fortunately, these instances were limited in large part due to the 53.34 cm (2 1/2 in.) flange width of the steel studs used to frame the exterior walls of this facility. This provided more tolerance when locating the studs and installing the fasteners compared to the 27.94 cm (1 1/2 in.) flange width found on more standard stud framing. Damage to the existing lamina due to overdriven fasteners or drilled holes was repaired prior to the cleaning of the existing lamina.

Because of the heavier than normal thickness of the steel studs (14 ga. steel as opposed to 16 or 18 ga. steel), the EIFS repair contractor had to establish on a trial-and-error basis how to install the fasteners to achieve adequate penetration of the fastener and plastic EIFS attachment washer without overdriving the screw which pulled through the plastic EIFS attachment washer. In order to establish a consistent base over which to apply the new lamina, the depressions at the fasteners were infilled with an application of noncementitious base coat troweled flush with the surrounding panel surfaces. Noncementitious base coat was then installed over the entire surface and returned into previously cleaned and prepared sealant joints. Standard weight mesh was troweled into the base coat to fully encapsulate the mesh.

At the shallow V-notch reveals, the detail mesh could not be installed since the build-up of base coat and mesh made the shallow reveals visually disappear. Therefore, the standard weight mesh stopped at the shoulders of the V-notch reveal, and only new noncementitious base coat and finish coat were applied in the reveal. Detail mesh and noncementitious base coat were applied at the sides and extended into the

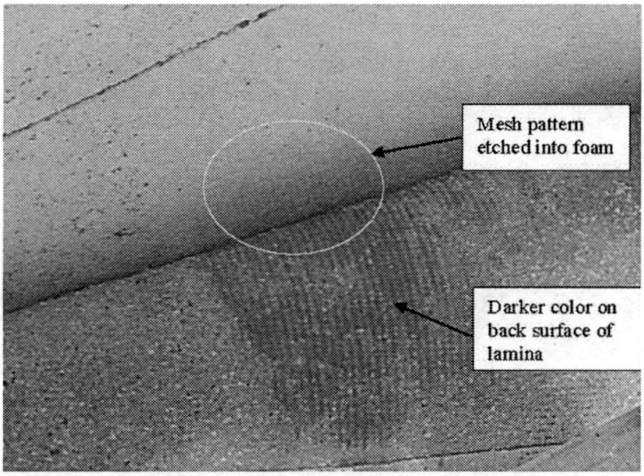

FIG. 12—*Lamina removed during repairs showed mesh pattern etched into the surface of the foam insulation and the corresponding back surface of the lamina distinctly darker.*

joints. After the base coat was cured, primer, backer rod, and sealant were installed. Once the sealant had sufficiently cured the finish coat was applied. Pull tests were performed during construction to verify that sufficient bond between the new and existing lamina was being achieved.

Staging of the Work

The work was sequenced to be completed in nine distinct stages, with scaffolding reaching the full height of the building in some areas. The staging was arranged to maximize use of the scaffolding and to efficiently schedule the work. Also, considerations were made to allow the owner continuous use of the building during the work. The scaffolding was screened for its full height to minimize visual impact during the repair work and to contain debris from the preparation of the EPS. The punchlist was performed as each stage of the work was completed, so corrections could be made before the scaffolding was dismantled and moved to a new location.

The repairs for the scaffolding ties were initially specified to include replacing the foam insulation and preparing the new base coat and finish coat with a typical EIFS patch. The contractor was concerned that these would be aesthetically objectionable. As an alternate repair, the contractor suggested cutting a 53.34 cm (2 1/2 in.) diameter hole in the wall around the scaffolding tie location and installing a 53.34 cm (2 1/2 in.) diameter plug cut from a prepared panel with 5.08 cm (2 in.) thick foam, polymer-based noncementitious base coat and mesh, and finish coat in the hole in the wall and adhering the plug to the sheathing with adhesive. The top surface of the plug would be flush with the top surface of the lamina on the wall, then the finish coat would be touched up around the perimeter of the plug. After this conceptual repair for the scaffold tie was evaluated, it was determined that the hole would be patched with a plug with base coat and mesh, and that the finish coat on the entire panel would be applied as the scaffolding is removed.

Pertinent Observations During Repair Work

During the course of the repair work which included the removal of the existing lamina, areas were observed where the mesh pattern was very clearly etched into the surface of the foam insulation. The back surface of the lamina positioned against these areas exhibited a distinctly darker color, as shown in Figs. 12 and 13.

The mesh pattern etched into the foam insulation was more pronounced in areas where the lamina exhibited pronounced debonding from the foam insulation (where the lamina could be described as "sagging" from the wall.) However, the mesh pattern was also etched into the foam in areas where the lamina visually appeared to be adhered to the foam.

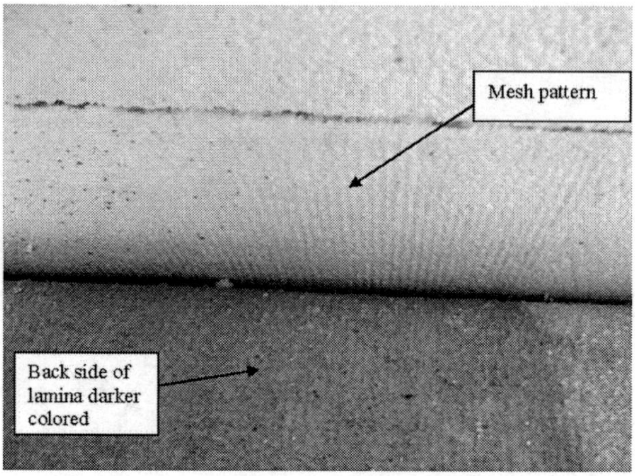

FIG. 13—*Lamina removed during repairs showed mesh pattern etched into the surface of the foam insulation and the corresponding back surface of the lamina distinctly darker.*

During the repair work, only small, isolated areas of mesh were observed to be in direct contact with the foam, which is contrary to what had been reported in investigations by others. In most locations, the mesh was observed to be fully encapsulated in the base coat and the base coat remained intact on the back surface of the lamina upon removal. This indicates that the delamination of the lamina due to plasticizer migration occurred in areas where the mesh was encapsulated in the base coat.

Melting or cavitation of the foam insulation was not observed during the repair work. The foam insulation could be prepared to receive the new lamina by rasping. Only one area was found where the surface of the foam insulation was observed to have a slightly brittle surface texture and was slightly yellowed; this coincided with pronounced delamination of the lamina and etching of the foam insulation surface, as shown in Fig. 14. Although the discoloration and texture difference were limited to the surface of the foam, the foam in this area was removed and replaced.

The observations made during the repair work, in conjunction with the laboratory testing, support the conclusion that the migration of plasticizer on the mesh used in the original construction caused the

FIG. 14—*In one area the surface of the foam insulation was observed to have a slightly brittle surface texture and was slightly yellowed.*

delamination of the lamina. Further, these findings do not support the conclusion of the initial investigation by others that the widespread delamination of the lamina was caused by melting of the foam insulation.

As described above, repairs were made in the areas where the lamina was removed and replaced utilizing mesh that did not contain the same plasticizer used in the original construction. The finish coat color of the new lamina matched that of the original construction, and a color coating was also used to reduce fading (and therefore maintain the relatively darker colors at the rust and dark beige areas). After three years of exposure to similar conditions, delamination of the lamina has not occurred. This further supports the conclusions that the migration of plasticizer on the mesh used in the original construction caused the delamination of the lamina, and that melting of the foam due to the color of the EIFS was not the cause of the delamination.

Conclusions

Based on the evaluation and testing, it was concluded that the failure of the lamina on this building was the result of the following:

1. The reinforcing mesh used in the original construction contains a plasticizer that can migrate and deteriorate the expanded polystyrene foam (EPS) with which it comes into contact. The deterioration of the EPS weakens the bond of the lamina to the EPS and therefore is the primary cause of the delamination of the lamina from the EPS.
2. The plasticizer migration can occur in the absence of elevated temperatures; however, elevated temperatures increase the rate of migration of the plasticizer and rate of reaction of the plasticizer with the EPS, and therefore accelerated the deterioration of the EPS and the resultant failure of the lamina.
3. The plasticizer migration can occur in all EIFS areas, regardless of the color. Since darker colors can experience higher base coat temperatures than lighter colors within a given set of environmental conditions, the rate of deterioration of the EPS and the resultant failure of the lamina is faster in areas where the EIFS surface color is darker.

Although initial repair recommendations were to completely remove and replace the existing lamina, it was ultimately determined that the existing lamina could be allowed to remain in some areas if the new lamina were adequately attached to the building. The new lamina system should utilize a mesh that does not have a plasticizer that is deleterious to the EPS.

Suggestions for Additional Testing and Evaluation

While the field observations made during the repair work and the subsequent performance of the repairs made to the building support the conclusions outlined above, additional testing and evaluation could likely provide additional evidence of this failure mechanism. For various reasons, this additional investigative work was not performed; however, the authors present these for the consideration of the reader:

1. In order to further explore the failure mechanisms of the delamination, in-place or laboratory testing could be performed on samples from the building that do not exhibit delamination by exposing the samples to varying levels of heat to the point of failure. In particular, the mechanism of plasticizer migration, the conditions of the lamina that influence plasticizer migration, and temperatures at which migration is initiated (if any) and progresses could be explored.
2. In order to further explore why the reveals did not exhibit widespread delamination, testing on the detail mesh could be performed to determine the type of plasticizer used on the detail mesh. Other conditions of the lamina in the reveals, such as base coat thickness, geometry, sun shielding, etc., could be explored further.

References

[1] *Guide to EIFS Construction*, EIFS Industry Members Association, 2000.

[2] ASTM Standard C578-04, 2004, "Standard Specification for Rigid, Cellular Polystyrene Thermal Insulation," *Annual Book of ASTM Standards*, ASTM International, West Conshohocken, PA, Vol. 04.06.

[3] ASTM Standard E2430-05, 2005, "Standard Specification for Expanded Polystyrene ("EPS") Thermal Insulation Boards For Use in Exterior Insulation and Finish Systems (EIFS)," *Annual Book of ASTM Standard*, ASTM International, West Conshohocken, PA, Vol. 04.12.

[4] Thomas, R., "Finish Line—Woe is Mesh," *Walls & Ceilings Magazine*, 1 July 2004.

[5] Thomas, R., "Finish Line—About Mesh," *Walls & Ceilings Magazine*, 8 July 2003.

[6] ASTM Standard E2134-01, 2001, "Standard Test Method for Evaluating the Tensile-Adhesion Performance of an Exterior Insulation and Finish System (EIFS)," *Annual Book ASTM of Standard*, ASTM International, West Conshohocken, PA, Vol. 04.12.

[7] ASTM Standard C1193-00, 2000, "Standard Guide for Use of Joint Sealants," *Annual Book of ASTM Standard*, ASTM International, West Conshohocken, PA, Vol. 04.07.

Journal of ASTM International, Vol. 4, No. 10
Paper ID JAI100871
Available online at www.astm.org

Eric K. Olson, P.E.[1] *and Judson A. Taylor,* AIA[2]

Retrofitting Barrier Exterior Insulation and Finish System: Methodology and Performance Assessment

ABSTRACT: Barrier-EIFS (exterior insulation and finish system) refers to EIFS with no internal drainage plane or through-wall flashings to collect and drain water that bypasses the EIFS's surface barrier. EIFS's surface barrier consists of a thin lamina and exposed joint sealants at transitions to elements such as windows, projecting decks, and intersecting parapets to prevent water infiltration through the surface barrier. Breaches in the surface barrier, however, can cause water leakage into the wall and building interior. In some cases reasonable reduction of leakage and preservation of the exterior walls may be accomplished by reducing barrier-EIFS's over-reliance on sealants, and by introducing watertight through-wall flashings at leakage-prone transitions and penetrations. The authors refer to this approach—providing flashings and localized drainage immediately around transitions and penetrations of wall elements while maintaining or improving the existing barrier-EIFS system away from such details—as "hybrid repairs." A case study of repairs on an existing high-rise barrier-EIFS-clad building shows the in-service performance of the hybrid repair methodology. The subject building had experienced water leakage at intersections of the EIFS and other building envelope elements, including windows, doors, balcony deck-to-wall transitions, and intersecting parapets. Repairs to correct water leakage focused on creating reliable concealed flashings at these details, along with more common barrier-enhancement repairs over the original EIFS, which was generally left in place. To help evaluate the hybrid repairs' effectiveness, the authors surveyed typical exterior repairs in place for 2–1/2 to 5–1/2 years and interior finishes in the areas of the repairs. The Authors interviewed residents in some units that had experienced chronic water intrusion prior to the repairs; residents commented that they have seen no water intrusion since the repairs. The survey and interviews with residents revealed no evidence or reports of reoccurring water intrusion, supporting the general effectiveness of the hybrid repairs on the subject building.

KEYWORDS: barrier-EIFS, EIFS repair, flashing, hybrid repair, threshold flashing, sill flashing, water intrusion

Introduction

Barrier-EIFS (exterior insulation and finish system)-clad buildings are commonplace throughout North America. Water leakage and resulting damage in walls constructed with barrier-EIFS have spawned general concern regarding its suitability for use on exterior building walls and the efficacy of sealant-based remedial approaches to mitigate water intrusion. Decisions regarding remediation of water leakage and damage and how to best prolong the life of a building's barrier-EIFS-clad exterior wall system have significant economic and functional consequences. Introducing drained flashings in existing barrier-EIFS, thus reducing its reliance on surface seals while retaining much of the in-place EIFS, can improve the durability and reliability of barrier-EIFS and yield a feasible solution that significantly reduces the risk of leakage and damage [1–7].

This case study presents a high-rise barrier-EIFS-clad condominium in Seattle, Washington. The building contains terraces and balcony decks accessed by sliding glass doors, parapets at balcony perimeters, and windows in punched openings (Fig. 1). This condominium, constructed in the early 1990s, began experiencing facade water leakage shortly after construction. Water intrusion was documented in over half of the 240 units. The water leakage led to rusted steel wall framing, mold growth, and damage to interior gypsum wallboard and other finishes. Leakage and resulting damage occurred primarily at wall bases inboard of balcony decks and adjacent to the sliding glass doors, beneath windows, and around details such as parapet terminations to walls. Much of the exterior wall system away from the leaking conditions, however, remained undamaged. After consultation, the homeowners selected a repair program

Manuscript received November 8, 2006; accepted for publication September 24, 2007; published online October 2007.
[1] Simpson Gumpertz & Heger Inc. 41 Seyon St.-Bldg. #1, Suite 500, Waltham, MA 02453.
[2] Simpson Gumpertz & Heger Inc. 1055 W. 7th Street, Los Angeles, CA 90017.

FIG. 1—*Building elevation. The case-study, high-rise barrier-EIFS-clad condominium has windows in punched openings and decks with and without perimeter parapets.*

that addressed leakage by introducing internal drainage features and flashings at leakage-prone areas while preserving much of the remaining EIFS. This case study describes the methodology, advantages, and limitations of this repair approach, and discusses the observed effectiveness of the in-place repairs.

Barrier-EIFS: Description and Discussion of Conventional Repair Approaches

Exterior insulation and finish system (EIFS) is a wall cladding consisting of a "lamina" (a thin layer of cement-based base coat, reinforced with glass fiber mesh and covered with a textured finish coat) applied over rigid insulation. The rigid insulation is applied over gypsum or wood exterior wall sheathing, in turn supported by wood or light-gauge framing. Barrier-EIFS (Fig. 2), unlike a drained and flashed wall system (Fig. 3), lacks an internal weather resistant barrier, a drainage plane, and flashings. The system relies instead on the barrier created by the lamina and joint sealants between the lamina and other wall elements to create a continuous surface barrier that blocks all water entry at the wall face. The barrier-EIFS facade system, as discussed herein, includes the EIFS wall panels plus other elements within the façade, such as windows and doors.

Barrier-EIFS, lacking internal drainage and flashings to collect and drain water that bypasses the surface seals, cannot tolerate face-seal or lamina surface breaches. Elements comprising the wall system, including the EIFS, exposed joint sealants, window and door frames, intersecting parapets and penetrations, therefore, must remain perfectly watertight. Water penetrating these wall elements becomes trapped within the wall, causing damage such as mold growth, deterioration of gypsum sheathing board, rusting of light-gauge steel framing and rotting of wood sheathing and framing. The historic performance of barrier-

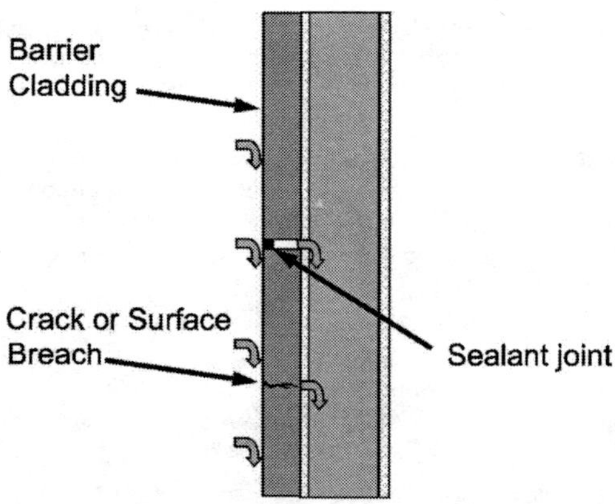

FIG. 2—*Surface-sealed barrier facade. The barrier facade relies on the unbroken facade surface and seals at joints to shed all water. Openings in the facade surface or joints allow water (arrows) to enter the interior of the wall.*

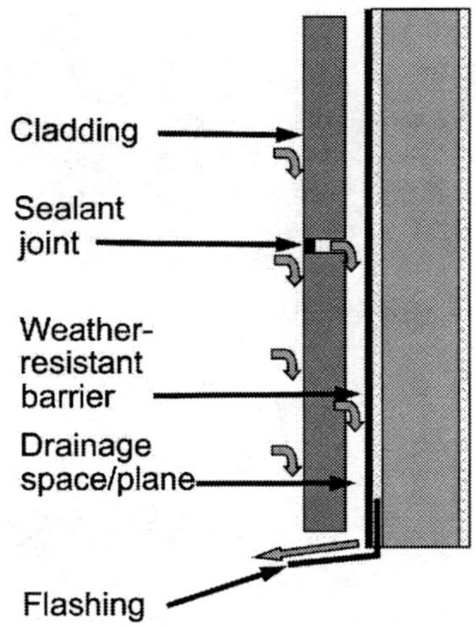

FIG. 3—*Drained facade. Water penetrating the facade surface or open seals (arrows) is drained by an internal weather resistant barrier and flashings at the base of the wall.*

EIFS shows that it is prone to water leakage through failed joint sealants between EIFS panels or between the EIFS and adjacent wall elements, cracked or punctured lamina, and unflashed or poorly flashed elements within the façade such as windows, intersecting walls and other penetrations.

One alternative in the remediation of barrier-EIFS water leakage is to maintain the EIFS's surface barrier by replacing failed sealants and by applying elastomeric coatings that help bridge cracks in the EIFS lamina. This barrier maintenance approach has limitations. It is often not possible to find and repair sealant breaches, cracks or damaged lamina before they allow significant leakage. This approach also does not attempt to address concealed leakage through components installed within the EIFS facade, such as windows and doors, as identification of such leakage typically requires invasive investigation and testing that are not feasible to conduct as a part of normal maintenance in an occupied building. Unaddressed, such leakage causes deterioration to gypsum sheathing, wall finishes, and other moisture-susceptible elements within the walls.

"Hybrid" Repair Approach

One approach to help overcome limitations of the barrier maintenance approach is to provide flashings and drained, dual-stage joint sealants in critical locations to collect and drain water from areas most vulnerable to water leakage. The resulting hybrid approach—providing flashings and localized drainage immediately around transitions and penetrations of wall elements while maintaining or improving performance of the existing barrier-EIFS panels away from such details—reduces the risks of leakage and resulting damage to levels that may be acceptable to certain building owners. This repair approach requires consideration of the following:

- Sufficient investigation has been undertaken to understand and identify systemic leakage paths and damaged wall elements so that a comprehensive repair scope can be developed. In practice, it is often impractical to find all concealed damage, so some local damage may remain in the wall in areas not exposed by the repairs.
- The hybrid approach requires selective demolition and reconstruction of EIFS to integrate new flashings with the existing wall. The entire façade will likely require recoating or finish coat renewal to provide an acceptable appearance match between patched and original surfaces.
- Existing barrier-EIFS wall panels retained by the hybrid repair scope may allow water leakage into the walls if the lamina cracks or become damaged following the implementation of repairs. Periodic inspection of the remaining EIFS can help mitigate, but cannot eliminate, these risks. The consequences of damage resulting from such water leakage are increased if oriented strand board (OSB)

FIG. 4—*Original condition of balcony decks. Poorly drained decks along EIFS walls and at sliding door thresholds contribute to failure of the sealant joint and water leakage along the base of the wall.*

sheathing or paper-faced gypsum sheathing, which deteriorate rapidly when exposed to moisture, are present.

- Repairs should only be considered for buildings with limited and localized exterior wall damage, and where the causes of leakage are linked to repetitive failure of discrete conditions within the facade. Extensive or system-wide damage to the EIFS or framing usually warrants re-cladding to restore damaged materials and wall structure.

The monetary savings generated by using barrier-EIFS with hybrid drainage features would be too small, and risks too high, to be well suited for new construction or re-cladding. We recommend a fully drained wall system for new construction or re-cladding.

Hybrid drainage repairs make the barrier EIFS facade system behave locally as a drainage-plane cladding by incorporating reliably watertight secondary barriers and flashings, capable of controlling and draining water, at vulnerable penetrations and transitions. While a complete, drained wall system cannot be created by local repairs to a surface-sealed wall system, the resulting hybrid wall system can provide substantial benefits of a drainage plane system—concealed and protected weatherproofing and less reliance on exposed sealants—by focusing repairs at critical conditions where these features are most needed.

Case Study: Causes of Leakage and Remedial Approaches

The case study building experienced widespread water intrusion caused by a limited number of repetitive conditions. Prior to implementation of the hybrid repairs, repairs to the surface seal system were attempted but did not correct the water intrusion problems. Investigation prior to developing the hybrid repair scope revealed that damage was generally limited to the glass-fiber-faced gypsum exterior wall sheathing, steel framing, and interior finishes adjacent to or directly beneath the water source. Away from leakage sources, the framing, exterior sheathing, and EIFS were in generally satisfactory condition. Deficiencies causing leakage and the repair approach to correct the conditions are discussed below.

Decks and EIFS Wall Base

The EIFS wall termination at balcony and terrace decks consisted of a sealant-filled joint, between the bottom edge of the EIFS and the urethane-based traffic coating covering the concrete decks. Poor drainage from the decks caused water to pond along the base of the wall (Fig. 4), exposing the EIFS lamina and sealant to water for prolonged periods and contributing to failure of joints between the deck and wall.

FIG. 5—*Wall base-to-deck flashing. Elastomeric base flashing at the deck-to-wall transition (arrow).*

Water penetrated these failed joint sealants between the deck and wall, damaging the wall framing and interior finishes. Repairs covered the transition between the deck surface and wall sheathing with an elastomeric base flashing, provided additional slope on the deck surface to drain water away from walls, and covered the new flashing and sloped fill with a traffic-bearing, waterproof coating (Figs. 5 and 6) to protect the new flashing. Metal "kick-out" flashings integrated with the elastomeric flashing and wall sheathing were provided where deck edges terminate into exterior walls (Fig. 7). Repairs also provided new metal pan flashing beneath doors following application of sloped fill to improve drainage away from the door. Wall base flashing and deck waterproof coatings were integrated with the pan to provide a continuous flashing along the base of the wall.

Windows

Windows had been installed with starter sills having end dams set in sealant. The starter sills did not extend beyond the face of the window to cover the sloped EIFS below; rather, the underside of the starter sill was sealed to the top edge of the EIFS panel below. Water penetrated window frame joints, window perimeter sealant joints, and openings in seals at starter sill end-dams. Repairs included installation of a metal sill flashing pan, integrated with rough opening jambs to capture jamb perimeter joint leakage (Figs. 8 and 9). The flashing pans have an extended sill to cover the sloped EIFS below the window sill and to eliminate the sill perimeter seal between the window frame and EIFS.

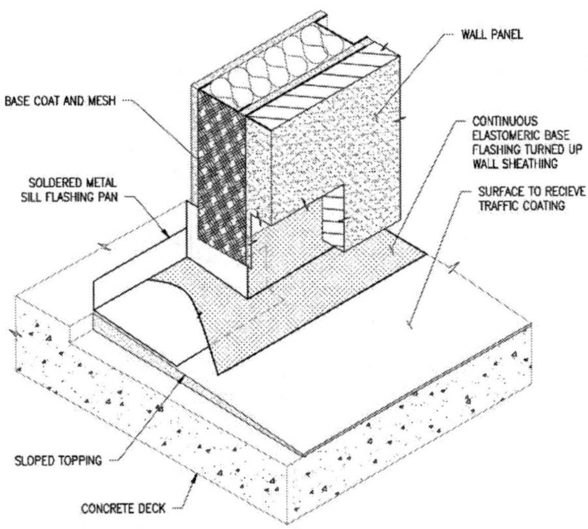

FIG. 6—*Wall base-to-deck flashing. Elastomeric base flashing at the deck-to-wall transition protects the base of the exterior wall and integrates the deck surface with door sill flashing.*

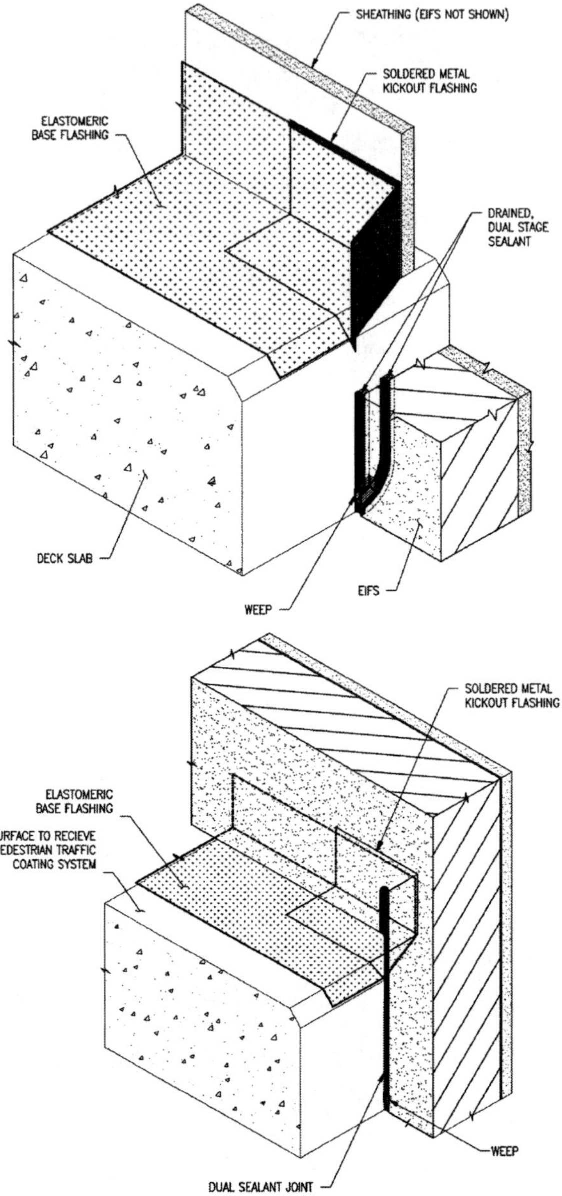

FIG. 7—*Deck edge termination to wall. Elastomeric base flashing along the wall base terminates inside a metal kickout flashing (top). The kick-out is terminated to the EIFS with a wept, dual-stage seal (bottom).*

Joint Sealants

Joint sealants around window and door perimeters, along the base of the EIFS panels at decks, and between EIFS panels, were adhered to the finish coat. Some joints had failed cohesively through the finish coat, allowing water intrusion; in other locations water penetrated the porous aggregate-based finish coat.

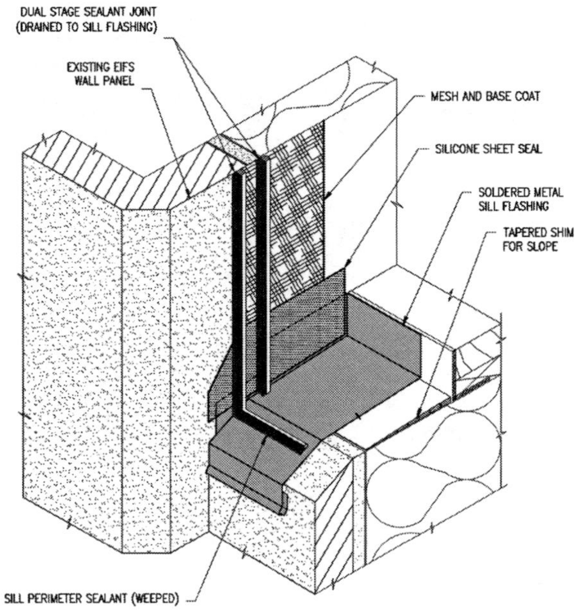

FIG. 8—*Window sill flashing. Window sill flashing pan drains frame leakage and covers and protects the sloped EIFS sill. A dual-stage jamb perimeter seal drains to the sill flashing.*

Repairs included removal of finish coat from joints and application of dual-stage joints, drained to flashing pans at the sills of windows and doors or drained to weeps at the base of vertical joints (Figs. 7 and 8).

Parapet Terminations to Walls

Parapets along deck edges are constructed of porous, split-faced concrete masonry block with a sheet metal coping. Concrete block and copings were sealed to the face of the EIFS. Water had penetrated the porous block and sealant at the coping-to-EIFS terminations at balcony and roof parapets. Repairs included

FIG. 9—*Window sill and parapet flashing. Installed window sill flashing (left arrow) and parapet-to-wall transition flashing (right arrow).*

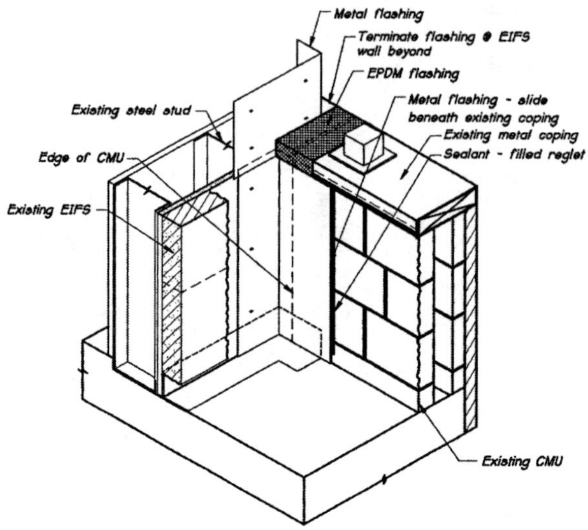

FIG. 10—*Parapet flashing to EIFS wall. Metal flashings integrated with the exterior sheathing and covered with new EIFS integrate the concrete block masonry parapets with the exterior wall.*

a metal flashing integrated with the wall sheathing, concrete block and deck base flashing (Figs. 9 and 10, and an elastomeric coating for concrete block surfaces. A drained sealant joint was installed within the gap between the new parapet flashing and EIFS.

Other Repairs

Conventional repairs to the case study building addressed cracking that occurred in some EIFS panels by application of new mesh, base coat and finish coat over existing cracked panels. The building received a silicone-based elastomeric coating to provide a consistent color match following installation of repairs and to prolong the service life of the existing finish coat. Split-faced concrete masonry unit (CMU) parapet surfaces received an elastomeric coating to minimize absorption of water into the block. Joints in metal parapet coping caps were covered with silicone sheet seals.

Assessment of Repair Effectiveness and Durability

The exterior envelope repairs at the case study building began with construction of a mock-up in a unit affected by severe water intrusion. The repair project started in early 2002 and concluded in late winter of 2004. We returned to the building in July 2006 to visually assess the condition of the repaired elements and to survey interior finishes in 18 units that had severe water-intrusion-related damage prior to repairs. We interviewed residents in 6 of the 18 units regarding reappearance of water or damage and interviewed property management personnel regarding reported leakage occurring since completion of the repairs. From these units, we also observed cladding components installed or modified during the repair process. We did not remove exterior cladding or interior wall finishes to observe concealed conditions.

Our evaluation of the repairs' effectiveness is based on the premise that ineffective repairs would have allowed water intrusion to recur in the locations seen prior to the repairs and that residents would have reported new leakage. We observed no recurrence of damage in the 18 units and building common areas we surveyed, though management personnel told us of new damage beneath a common area skylight from sources unrelated to the repairs. Interviewed residents and property management personnel have reported no other recurring or new leakage.

We reviewed the condition of exterior wall sealants, elastomeric coating, deck waterproofing and flashing at twenty units (including the 18 units having damage prior to the repairs) and at accessible terrace decks and the roof. Our summary of observations of the present condition of exterior elements is as follows:

- Surface of the EIFS lamina painted with a silicone-based elastomeric coating: The coating remains well adhered with no areas of peeled or blistered coating. The coating surface has, however, accumulated some dirt; this can be expected with silicone-based elastomeric coatings. The dirt stains are most pronounced where water runs down coated surfaces (e.g., from sloped metal roof areas).
- Previously cracked EIFS panels refinished with new base coat, mesh, and textured acrylic finish coat: No cracking or peeling of the refinished EIFS panels.
- Low-modulus silicone sealant between EIFS panels and at window and door perimeters: Sealants remain intact and adhered to substrates.
- Preformed silicone sheet seals at parapet coping joints: Isolated tears in sealant sheet at several sheet metal coping joints.

Conclusion

The localized repair and retrofitting of an existing EIFS-clad facade to provide flashings and drainage at leakage-prone sealant joints and details can substantially reduce water leakage into the wall, thereby extending the façade's service life and reducing the risk of water intrusion to acceptable levels (as determined by the building owner). In cases where leakage and consequential damage are limited and can be reasonably well located, the approach provides a cost-effective solution to leakage and durability problems with the barrier EIFS-clad facades. Risks of this approach include isolated, remaining damage concealed within the wall system and future leakage arising from openings such as cracks or punctures in remaining barrier-EIFS panels.

The hybrid repair approach is a compromise between the competing goals of eliminating all wall leakage and reducing remedial costs and disruption associated with repairs to barrier-EIFS-clad walls. Ultimately, the building owner must select the repair scope, based upon an analysis of their level of acceptable risks and rewards associated with each option.

References

[1] Bronski, M. B., and Ruggiero, S. S., "Exterior Insulation Finish Systems (EIFS) Use in Wood-Framed Construction: Design Concepts to Avoid Common Moisture Intrusion Problems," *J. Test. Eval.*, Vol. 28, No. (4), 2000, pp. 290–300.

[2] French, W. R., "The Practical Use and Potential Limitations of Exterior Insulation and Finish System Materials as an Exterior Building Envelope," *Water in Exterior Building Walls: Problems and Solutions*, ASTM Standard STP 1107, T. A. Schwartz, Ed., ASTM International, Philadelphia, PA, 1991.

[3] Nelson, P. E., and Waltz, M. E., Jr., "EIFS—Surface Sealed Wall Systems that Need Flashings," *Exterior Insulation Finish Systems (EIFS): Materials, Properties and Performance*, ASTM STP 1269, P. E. Nelson and R. E. Kroll, Eds., ASTM International, Philadelphia, PA, 1996.

[4] Olson, E. K., and Taylor, J. A., "Strategies for Prolonging the Service Life of Barrier-EIFS," *The Construction Specifier*, Vol. 56, No. (12), 2003, pp. 53–61.

[5] Piper, R., "Troubles with Synthetic Stucco," *New England Builder*, Vol. 6(9), June 1988, pp. 34–38.

[6] Piper, R., "Hidden Damage Likely on an EIFS Failure," *Condo Media*, June 1994, pp. 35–36.

[7] Ruggiero, S. S., and Myers, J. C., "Design and Construction of Watertight Exterior Building Walls," *Water in Exterior Building Walls: Problems and Solutions*, ASTM Standard STP 1107, T. A. Schwartz, Ed., ASTM International, Philadelphia, PA, 1991.

Author Index

Subject Index